7th Biennial Conference of the Interdisciplinary Society for Quantitative Research in Music and Medicine (ISQRMM 2023)

Published in the Journal of the Interdisciplinary Society for Quantitative Research in Music and Medicine

Volume No. 7 (2023)

Online
3-4 June 2023

Editors:

Kent P. Nelson
David O. Akombo
Adina Dabija

ISBN: 979-8-3313-1335-7

Printed from e-media with permission by:

Curran Associates, Inc.
57 Morehouse Lane
Red Hook, NY 12571

Some format issues inherent in the e-media version may also appear in this print version.

ISBN: 979-8-3313-1335-7

Printed by Curran Associates, Inc. (2025)

For permission requests, please contact ISQRMM
at the address below.

ISQRMM
C/O David O. Akombo, Ph.D.
Dean, Faculty of Culture, Creative and Performing Arts
P.O. Box 64
Bridgetown BB11000
Barbados, West Indies

http://isqrmm.org

The *Journal of the Interdisciplinary Society for Research in Music and Medicine* is a peer-reviewed journal presenting scholarly research into music's connection and application in medicine. Articles are submitted from scholars who have presented at the society's biannual conference, as well as from the field. Inquiries regarding submission should be directed to contact@isqrmm.com

Additional copies of this publication are available from:

Curran Associates, Inc.
57 Morehouse Lane
Red Hook, NY 12571 USA
Phone: 845-758-0400
Fax: 845-758-2634
Email: curran@proceedings.com
Web: www.proceedings.com

Journal of the Interdisciplinary Society for Quantitative Research in Music and Medicine
Volume No. 7 (2023)

Table of Contents

President's Message

It is with great pride and excitement that we publish this edition of the *Journal of the Interdisciplinary Society for Quantitative Research in Music and Medicine* (JISQRMM). This journal serves as a vital bridge between the fields of music, medicine, and both quantitative and qualitative research, fostering a deeper understanding of how music and art influence human health and well-being. The relationship between music and medicine—manifested in remarkable ways from Ancient Greece to the 18th century and inspiring scholarship on ancient civilizations—has seldom been thoroughly examined through the lens of contemporary scientific models. Furthermore, there is a notable lack of research exploring global cultures that employ music as a form of non-pharmacological medicine, an area deserving of deeper study. The *Journal of the Interdisciplinary Society for Quantitative Research in Music and Medicine* (JISQRMM) is dedicated to advancing knowledge at the intersection of these disciplines. Its mission includes promoting interdisciplinary collaboration, supporting evidence-based research, encouraging the integration of music into healthcare, and providing a platform for disseminating innovative findings, methodologies, and theoretical perspectives.

Currently, much of the information on the culturally diverse applications of music as medicine comes from anecdotal accounts, which document the experiences of individual music healers. While quantitative studies in this field are invaluable, the anecdotal evidence recorded in treatises on healing rituals provides significant historical credibility to music's role as a healing agent. The *Journal of the Interdisciplinary Society for Quantitative Research in Music and Medicine* welcomes both qualitative and quantitative research methodologies and encourages interdisciplinary approaches. Authors are urged to employ research methods that are clearly defined, well-articulated, and relevant to their objectives.

The journal also invites submissions based on master's or doctoral theses with significant impact or historical relevance, especially those focusing on qualitative research in music and healing. By featuring cutting-edge research that balances diverse methodologies, the journal appeals to an interdisciplinary readership. Its eclectic models for presenting findings ensure that the content remains both scholarly and accessible to readers across various fields of expertise. We take immense pride in the ISQRMM community—its passionate enthusiasts and distinguished scholars—who together comprise our global network of researchers.

We are proud of ISQRMM community, enthusiasts, and erudite scholars that make up our global research clientele.

Musically Yours,

Professor David O. Akombo, PhD, M.A., M.Mus., B.Ed.

President and Co-Founder

Co-Founders

David O. Akombo, Ph.D., M.A., M.Mus., B.Ed.

Joseph Ruszkowski, Ph.D.

Estate of Arthur Harvey, DMA (Professor of Music and Psychology, University of Hawaii)

Estate of Jorge Camara, M.D. (Professor of Ophthalmology, University of Hawaii School of Medicine)

Board Members

David O. Akombo, Ph.D. (Professor of Music Education and Interdisciplinary Musicology)

Kent Nelson, DMA, (Independent Scholar)

Adina Dabija, LAc, MSTOM, Dipl OM

Juliana Azoubel, M.A., (PhD Candidate) Professor of Dance, Federal University of Minas Gerais, Brazil

Editor's Note

Dear Readers, Researchers, and Members of the ISQRMM Board,

Welcome to the 7th Biennial Conference of the *Interdisciplinary Society for Quantitative Research in Music and Medicine* (ISQRMM) Journal. I trust you will find the contents of this journal both comprehensive and informative. The research presented by top scholars explores historical contexts while offering advanced and innovative applications of music and the arts for promoting a healthier future.

Speaking of health and the future, ISQRMM is stronger than ever, as reflected in this journal's contents. Together, we have weathered the challenges of the pandemic. Now, the research featured in ISQRMM journals may guide us toward better, healthier, and more fulfilling lives. The strength of ISQRMM lies in the diversity of its members' interests and approaches. This diversity is once again reflected in the articles presented here. From the application of ancient Greek musical traditions for music therapists, to sound healing, concentration enhancement through music, harmonization of the body via deep breathing, and harp therapy, music offers numerous pathways to promote healing and physical and mental well-being.

While the healing properties of music have been recognized for centuries, only recently have researchers in Western cultures begun systematically exploring its benefits beyond aesthetic enjoyment. Music is magical, transforming organized tonal air vibrations into sounds that deeply impact both body and mind. Formal, systematic research on the health benefits of music is essential for advancing our understanding of this phenomenon. I invite you to read and reflect on the contents of this 2023 ISQRMM Journal. Consider how the authors' insights might enhance your life, delve deeper into the topics, and explore the references provided in each article for a wealth of further information.

One last word, the contributors of this journal enjoy hearing from you. Your questions, insights, or even a quick communication via email that you have read their articles is encouraging to these authors. We are all a community that needs the exchange of ideas and experiences to assist in the fascinating field of music and health. Many of the contributors to the journal have included their contact information as seen at the end of the journal beginning on page 162, the Contributors section. Furthermore, it is there that you can read of the extensive background, education, and further research that these authors have done.

It has been a pleasure serving as the editor of the 2025 Journal for the ISQRMM, and I hope you find the articles enlightening and inspiring.

Kent P. Nelson, DMA
Editor and Board Member

Co-editors:
David O. Akombo PhD
Adina Dabija, LAC, MSTOM, Dipl OM

The *Journal of the Interdisciplinary Society for Research in Music and Medicine* is a peer-reviewed publication dedicated to presenting scholarly research on the connection between music and medicine and its applications. Articles are submitted by scholars who have presented at the society's biannual conference, as well as by experts in the field. Inquiries regarding submission should be directed to contact@isqrmm.com

Effects of Piano Practice on Anxiety in Teenagers and Young Adults

David O. Akombo

Abstract

Context: students at the university may experience some kind of anxiety brought about by varied factors such as academic pressure, responsibilities, and social adjustment in a new environment. University students' lives are often demanding and thus exposing them to anxiety. **Objective:** To determine whether piano practice influences anxiety levels in teenagers and young adults. **Design:** In this repeated measures design, twenty-five (25) undergraduate students above 18 years of age participated in the study (n = 25). **Setting:** Public institution of higher learning in Florida. **Intervention:** Participants who were enrolled in studio piano from an institution of higher learning in Florida were enrolled in the study. **Hypothesis:** Pre- and post-intervention measurements of anxiety showed that piano practice in the studio had a significant effect on college students' anxiety levels compared to the baseline.

Keywords: Anxiety, Piano, Music Practice, Stress

Introduction

Anxiety is differentiated into two distinct categories; (a) State Anxiety, a transitory emotional reaction to the individual's perception of a threatening or dangerous situation; and (b) Trait Anxiety, a relatively stable tendency to interpret situations as threatening or dangerous and to react to them with anxiety (Basco & Olea, 2013). It is certainly not unusual for a college student who has worked long and hard in the week including moments of overnight class discussions and term papers to exhibit some form of anxiety. Music and anxiety have been studied for more than three decades both as intervention through music therapy applications or as investigative studies on cognitive and other outcomes. For the most part of the last three decades, scholars have had great interest in the use of music to control anxiety. Alleviation of stress anxiety has been addressed by Hammer (1996), who studied the relationship between music therapy and participants' perceived stress level. Participants in the Hammer study joined in music therapy sessions that included relaxation techniques and guided imagery. The State and Trait Anxiety Inventory (STAI) was administered before and after the music therapy sessions. In the experimental group, there was a slight reduction in STAI levels and a perceived decline in stress levels after the music therapy sessions. When considering the results of the study, Hammer concluded music could be effectively used to reduce stress anxiety.

In another experiment, Brennan & Charnetski (2000) studied the effect of music in reducing stress anxiety by bringing music into the workplace. Their experiment studied the effect music has on self-reported stress levels and the immune system. The researchers obtained baseline levels of self-reported stress and measured the presence of salivary immunoglobin A (S-IgA) in the blood of the participants. S-IgA usually is an indicator of immune system activity. This pretest and posttest experimental design was designed in such a way that after playing smooth jazz in the participants' high-stress newsroom workplace, they measured the participants' levels of self-reported stress and the levels of S-IgA in the blood. The researchers were able to conclude that the playing of recorded smooth jazz music resulted in a reduction on the participants' perceived level of stress anxiety.

Researchers have continued to show the effect of music on human health. Brownley et al. (1995) carried out a research study on the influence of music on physiological and affective exercise responses in eight trained and eight untrained runners under three music conditions

https://doi.org/10.52202/078697-0001

("no," "sedative," and "fast") during low, moderate, and high intensity exercise. Repeated results of this study revealed increased respiratory frequency (greater excitation) during fast music as compared to the no music and sedative music conditions.

In a study on music and labor pain, Geden et al., (1989) carried out two studies to examine the effects of music on analogue labor pain using volunteer nulliparous subjects who were randomly assigned to treatment groups. Assessments of the treatments were made in a 1-hour session involving twenty 80-second exposures to a laboratory pain stimulus patterned to resemble labor contractions. In the first experiment, it was hypothesized that subjects listening to easy-listening music would report lower pain ratings and cardiovascular responses than subjects listening to rock music, self-selected music, or a dissertation (placebo-attention) as subjects in a no-treatment control group. Significant group effects were found between each listening and between all other groups. Subjects spontaneously reported using imagery as a pain reduction technique. In the second study a combination of music and imagery was examined by randomly assigning subjects to one of five groups: self-generated imagery with music (SIM), guided imagery with music (GIM), self-generated imagery without music (SI), guided imagery without music (GI), or no-treatment control. Again, significant differences were obtained between the with music and the without music for all groups.

In another study on the effect of music on stress anxiety levels, Migneault et al. (2004) examined the effect of music on the neurohormonal stress response to surgery under general anesthesia. Their study found that several pharmacological interventions reduce preoperative stress hormone release during surgery under general anesthesia. They also studied listening to music and therapeutic suggestions where they found a positive effect on postoperative recovery and the need for analgesia. In this study, they evaluated the effect of listening to music under general anesthesia on the neurohormonal response to surgical stress as measured by appropriate medical procedures. Music therefore had an effect even under anesthesia.

In their research where they attempted to the answer the question: "Does Singing Promote Well-Being?", Grape et al., (2004) carried out an empirical study of professional and amateur singers during a singing lesson. This study explored the possible beneficial effects of singing on an individual's well-being during a singing lesson. Eight amateur (2 males, 6 females, ages 28-53 years) and eight professional (4 males, 4 females, ages 26-49 years) singers who had been attending singing lessons for at least six months were included. Five visual analogue scales

https://doi.org/10.52202/078697-0001

(VAS, sad-joyful, anxious-calm, worried-elated, listless-energetic, and tense-relaxed) were scored before and after the lesson. In addition, a semi-structured interview was performed. Heart rate variability analyses showed significant changes over time in the two groups for total power, and low and high frequency power. Power increased during singing in professionals, whereas there were no changes in amateurs. This indicates an ability to retain more "heart-brain connection," i.e., more cardio-physiological fitness for singing in professional singers, compared to amateur singers. In another study on anxiety, Yilmaz et al. (2003) carried out a study to find out if music decreases anxiety and provides sedation in a procedure known as extracorporeal shock wave lithotripsy (ESWL). In this procedure, sound waves are used to break a kidney stone into small pieces that can more easily pass into the bladder. The researchers of this study revealed that listening to music by patients during (ESWL) treatment is a feasible and convenient alternative to sedatives and anxiolytics.

Winter et al. (1994) observed the well-known fact that many patients become stressed and anxious prior to and after surgery and that one means of helping reduce anxiety in patients is to incorporate music in the Surgical Holding Area. In their study, they divided their subjects into two groups: one group of patients listened to music while a second group did not. The researchers observed that patients who listened to music while in the Surgical Holding Area had significantly less stress and anxiety than did those who did not listen to music. Both groups spent similar lengths of time in the Surgical Holding Area. Winter et al. concluded that patients preferred the musical intervention since they experienced less anxiety.

In another study, researchers at the Bryan Memorial Hospital, Lincoln, investigated the influence of music therapy on mood and anxiety of patients undergoing heart surgery (Miers, 2018). Ninety-six patients who underwent elective, heart bypass surgery at the cardiovascular intensive care and progressive care units of a Midwestern community hospital participated in the study. Data relating to anxiety and mood was obtained through blood pressure and heart rate using the Spielberger's state-trait anxiety inventory (STAI). Through their study, the researchers revealed that the patients' mood ratings showed significant improvement in mood among those patients who were in the "music therapy" group after the second intervention. There were also significant main effects over time for heart rate and systolic and diastolic blood pressure in the music therapy group, which indicated a generalized physiologic relaxation response. Reduced

https://doi.org/10.52202/078697-0001

anxiety and improved mood were observed in all three groups, and the researchers noted that all the interventions resulted in a generalized relaxation.

Augustin & Hains (1996) examined the role that music therapy might play in a postoperative setting for ambulatory patients. In the Day Surgery Unit of St. Mary's Hospital, Mequon, Wisconsin, they investigated forty-two ambulatory surgery patients where they assigned them to either an experimental group to receive music therapy along with the standard preoperative instructions or a control group to receive the standard preoperative instruction alone. The results revealed that the patients in the experimental group showed significantly lower heart rates compared to the patients in the control group. The experimental group also showed greater improvements in blood pressure and respiration rate. The researchers concluded that music therapy offers demonstrable benefits for ambulatory surgery patients and recommend that the patients be offered music as an effective option to help alleviate postoperative anxiety.

The effect of music in the alleviation of physiological conditions such as anxiety is a historical phenomenon. This practice is evidenced by several mythological reflections, which have explained the origins of music in healing. One belief derives the use of music in healing from the Old Testament in the Bible (circ. 1445 B.C.) where David played his harp to Saul for curative purpose (Benenzon, 1997; Moreno, 1985; Darrow et al., 1985). Some attribute the rise of music in healing to the life in Stone Age, which is saturated with the practice of using music for health. The Stone Age cultures did not view music as a self-contained experience apart from life. The connection between music and healing has come a long way from its primitive stage; but this connection has been retained with the human existence. In this regard, we may speak of music as a universal agent that transcends all time and space, barriers withstanding in alleviating suffering such as that caused by pain and anxiety. Its use as a healing agent has been shaped over thousands of years through human experience.

Method

All subjects were either majoring in piano or studying piano as one of their secondary instruments. The subjects practiced music in the practice rooms in a university in the USA. The selection process was based on opportunity sampling. Thus, there was no random sampling but those who wished to participate in the study were recruited and given the State Trait Anxiety Inventory questionnaire to fill 10 minutes before practice and immediately after practice. The criteria for subject participation included students being able to practice music on piano for at least half an hour in order to participate in the study. The subjects also had to be at least in their intermediate level of musicianship, and they had to be over the age of 18. The subjects had to be able to read and understand English (as only the English version of the STAIC scale was used). Prior to participation in the study, subjects were informed of the study's purposes. If the subjects agreed to participate, they reviewed and signed a consensus form. The study was approved by the Institutional Review Board (IRB).

Materials

Informed Consent forms were used which informed the participants of the purpose of the study and explained that the participants would be filling out two surveys. The tool used to measure anxiety was The State-Trait Anxiety Inventory (STAI), which was invented by Spielberger, Gorusch, and Lushene in 1957. Forms C-1 and C-2 of the State Trait Anxiety Inventory (STAI) were used to measure the participants' current level of stress anxiety on two occasions. According to Spielberger et al., (1983), state anxiety reflects a transitory emotional state or condition of the human organism that is characterized by subjective, consciously perceived feelings of tension and apprehension, and heightened autonomic nervous system activity. State anxiety may fluctuate over time and can vary in intensity. In contrast, trait anxiety denotes relatively stable individual differences in anxiety proneness and refers to a general tendency to respond with anxiety to perceived threats in the environment.

Design and Procedure

The study used a single-variable pre-test/ post-test within groups design. The independent variable included music and practice time while the dependent variable was anxiety. STAIC forms C-1 and C-2 at least 10 minutes before their practice. Subjects had to practice at least for 30 minutes without interruption. The subjects had to complete the STAIC forms C-1 and C-2 one more time after the practice session.

Results

The mean pretest state anxiety score was 34.8, while the mean post-test state anxiety score was 33.44. The mean pretest trait anxiety score was 34.48 while the mean posttest Trait anxiety score was 34.64. I hypothesized that post measures of stress anxiety for state would differ significantly with the sexes with female being the least stressed and the male group being the most stressed.

https://doi.org/10.52202/078697-0001

This hypothesis was not supported by the results obtained (male M= 33.2, female M= 31.2) However, there was a correlation relationship between the pre-test and the post-test STAI scores (r = .903.)

Discussion

The researcher hypothesized that same levels of stress anxiety would be reported after the music practice in every participant. It was also hypothesized that females would not show any significant differences in the pretest and posttest anxiety levels from those of males. The present data were not found to be significant, and the null hypothesis cannot be rejected. The mean scores went down for both sexes on the posttest of the STAI, but it did not reach significance. Even though there were very minimal controls to this study, physiological measures of heart rate, respiration rate, blood pressure, skin temperature, EEG, and neurohormone levels would all have been beneficial and would provide more comprehensive assessment. In this study, one respondent indicated that he often felt better after his music practice. Another respondent said that he normally practices for
long hours and that the longer he played the better he felt.

Some possible reasons why the data is not congruent with past research could be the length of time that the participants were exposed to music and the type of music they played. The other reason could be that the participants played music only for a short period of time. In this experiment, the participants were exposed to music for a short period of time, about thirty minutes, whereas in much of the past research, participants were exposed to the music for longer periods of time (Hammer, 1996), and in some cases an entire day (Brennan & Charnetski, 2000). Participants were exposed to music for only a short period of time in this current study in order to increase the external validity of the study. People often cannot practice for long periods during the day, and maybe only while they are driving or exercising. In an effort to increase the external validity, the internal validity was compromised and that could have contributed to the data not being significant.

Another possible reason could be the use of the STAI. Some of the past research used more objective measurements such as salivary recordings and measuring the nervous system activity (Brennan & Charnetski., 2000; McCraty et al., 1998). Because the participants were administered two similar surveys within a short period of time, their answers on Form C-2 might have been influenced by their answers on Form C-1, an idea supported by the correlation

https://doi.org/10.52202/078697-0001

between the pre-test and post-test scores. The high positive correlation (r= .903) between participants' answers on the two surveys indicated that their answers on the first survey would be the best predictor of their answers on the second survey, rather than the type of music they practiced. The study did discover some significant relationships between the sexes and how their anxieties fluctuated with music practice. General Linear Model indicated that the State anxiety overall decreased across time. There was a trend for the female participants to show a larger decrease than the male participants, but it does not reach significance suggesting that sex might not have as much of an effect as was originally hypothesized. Because the participants were only exposed to the music for a short time, the results could be indicative of the fact that participants must be exposed to music for a much longer time for the music to significantly affect stress anxiety levels. Therefore, practicing music will not significantly lower stress anxiety levels if people only practice for short periods of time. Also, music therapy was not incorporated into this study. Music therapy, such as the guided imagery used in Hammer's (1996) experiment or the positively induced emotional state used in McCraty et al.'s (1998) experiment, compels the participant to become actively involved in reducing their stress levels while playing music. This study could have implications for people who passively practice specific types of music intended to relieve stress, for short amounts of time, and think that they are lowering their stress levels. The results of this study may indicate that this belief is not valid. People may need to practice music throughout the day and incorporate some type of music therapy, such as guided imagery, into their music experience to gain any of the possible benefits.

Because this study did not discover a significant relationship between music practice and stress anxiety, it could be viewed as weakening the prevailing view that music does indeed affect stress anxiety. But rather than negate the bulk of the past research (Hammer, 1996; Brennan & Charnetski, 2000; Standing & Stace, 1980) that does support the hypothesis that music does affect stress anxiety, the results of this study could have greater implications for the perceived strength of the relationship between stress and music practice. Since music was only practiced for a short period of time, the relationship between music practice and stress anxiety would have to be strong for the results to be significant. A relationship could exist, but it might be too weak to have been demonstrated in this study.

For future studies, researchers should expose the participants to music practice for longer time periods so that the music could have a greater effect. Also, using biological methods to

https://doi.org/10.52202/078697-0001

measure stress anxiety such as physiological measures of heart rate, respiration rate, blood pressure, skin temperature, EEG, and neurohormone levels would be preferable. These methods are more expensive but provide more objective and reliable data.

It would be interesting to do further studies on the effects of forced music practice on stress and anxiety. Future studies could also focus on the effects of music practice on stress anxiety and the physical health of the participant and additionally focus on the strength of the relationship. The results of future studies could be used to implement simple and inexpensive ways to alleviate stress anxiety in the general population by providing music practice rooms in hospitals, schools, and workplaces.

References

Augustin, P.& Hains, A. A. (1996). Effect of music on ambulatory surgery patients' preoperative anxiety. *AORN Journal 63*(4) 750, 753-758. doi: org/10.1016/s0001-2092(06)63126-8

Basco, R. E., & Olea, M. T. (2013). Correlation between Anxiety Level and Academic Performance of BS Biology Freshmen Students. *International* Journal *of Educational Research and Technology, 4*(1), 97-103.

Benenzon, R. (1997). *Music therapy theory and manual: Contributions to the knowledge of nonverbal contexts*. Springfield, IL: Charles C. Thomas.

Brennan, F. X., & Charnetski, C. J. (2000). Stress and immune system function in a newspaper's newsroom. *Psychological Reports, 87*(1), 218-222. doi.org/10.2466/pr0.2000.87.1.218

Brownley, K. A., McMurray, R. G., & Hackney, A. C. (1995). Effects of music on physiological and affective responses to graded treadmill exercise in trained and untrained runners. *International Journal of Psychophysiology: Official Journal of the International Organization of Psychophysiology, 19*(3), 193–201. doi: org/10.1016/0167-8760(95)00007-f

Darrow, A. A., Gibbons, A. C., & Heller, G. N. (1985). Music therapy past, present, and future. *The American Music Teacher, September/October,* 18-20.

Geden, E. A., Lower, M., Beattie, S., & Beck, N. (1989). Effects of music and imagery on physiologic and self-report of analogued labor pain. *Nursing Research, 38*(1) 37-41.

Grape, C., Sandgren, M., Hansson, L. O., Ericson, M., & Theorell, T. (2003). Does singing promote well-being?: An empirical study of professional and amateur singers during a singing lesson. *Integrative Physiological & Behavioral Science: The Official Journal of the Pavlovian Society, 38*(1), 65–74. doi: org/10.1007/BF02734261

Hammer, S. E. (1996). The effects of guided imagery through music on state and trait anxiety. *Journal of Music Therapy, 33*(1), 47-70. doi: org/10.1093/jmt/33.1.47

McCraty, R., Barrios-Choplin, B., Atkinson, M., Tomasino, D. (1998). The effects of different types of music on mood, tension, and mental clarity. *Alternative Therapies in Health and Medicine 4*(1), 75-84.

Miers, D. (2018). Music for Mental Health. Accessed June 13, 2024 from https://www.bryanhealth.com/about-bryan-health/news/2018/music-for-mental-health--/

https://doi.org/10.52202/078697-0001

Migneault, B., Girard, F., Albert, C., Chouinard, P., Boudreault, D., Provencher, D., Todorov, A., Ruel, M., & Girard, D. C. (2004). The effect of music on the neurohormonal stress response to surgery under general anesthesia. *Anesthesia & Analgesia, 98*(2), 527-32. doi: 10.1213/01.ANE.0000096182.70239.23

Moreno, J. (1995). Ethnomusic therapy: An Interdisciplinary approach to music and healing. *The Arts in Psychotherapy, 22*(4), 329-338.

Spielberger, C., Gorusch, R. L., Lushene, R., Vagg, P. R., & Jacobs, G. A. (1983). *Manual for the State-trait Anxiety Inventory*. Redwood City, CA: Consulting Psychologists Press.

Standing, L., & Stace, G. (1980). The effects of environmental noise on anxiety level. *Journal of General Psychology, 103*(2d Half), 263-72. doi: 10.1080/00221309.1980.9921007

Winter, M. J., Paskin, S., & Baker, T. (1994). Music reduces stress and anxiety of patients in the surgical holding area. *Journal of Post Anesthesia Nursing, 9*(6), 340-343.

Yilmaz, E., Ozcan, S., Basar, M., Basar, H., Batislam, E., & Ferh, at M. (2003). Music decreases anxiety and provides sedation in extracorporeal shock wave lithotripsy. *Urology, 61*(2), 282-286. doi.org/10.1016/s0090-4295(02)02375-0

The Orphic Path to Divine Union: "Theon Therapeian" and Ritual Music in Contemporary Greek Mysteries

Antonios Chrysovergis

Abstract

Presented in this paper, I explore the value of "Theon Therapeian," an ancient Greek therapeutic idea that emphasizes "Service to the Gods" and "Divine Worship" through ritual music. In this study, I focus on the Orphic tradition and religion in the context of contemporary Greek Mysteries. I am seeking to reveal the transformative power of Divine Worship and ritual music in achieving unity between the human soul, nature, and the divine. My research provides a brief overview of Orphic beliefs as practiced today, along with examples of contemporary Orphic rituals and musical illustrations. Ultimately, my aim is to revive and elevate ancient practices, inspiring interest in the transformative and therapeutic power of both ancient and contemporary Greek Mysteries and esotericism.

Key Words: Orphic, Greek, Mysteries, Music, Therapy

https://doi.org/10.52202/078697-0002

Introduction

The Orphic tradition, a mystery tradition originating from ancient Greece, draws deeply from the life and epoch of Orpheus – a revered figure from remote antiquity. This tradition profoundly intertwined elements of music, religion, mythology, science, and philosophy, among others. In this paper, I aim to highlight the Greek understanding of the Orphic Mysteries and tradition, as well as the identity of Orpheus, his followers, and the traditions and practices associated with him. Most ancient authors, along with his ancient and contemporary followers, regarded Orpheus as a historical figure rather than solely a character of mythology.

According to a popular version of local tradition, Orpheus, a Greek prince of great significance, was born in a cave on Mt. Olympus near Leibethra, an ancient city in the region of Pieria. At the end of the Orphic Argonautica, one of the ancient texts to which literary authority is attributed to Orpheus himself, Orpheus pays tribute to his birthplace: "Thus, quickly departing, I pressed on to snowy Thrace, in the region of Libethra, in my homeland, entering the famous cave where my mother gave birth to me on the bed of the brave Oeagrus" (trans. 2011). Leibethra was also Orpheus' burial place according to tradition. This region, now located in central Macedonia with Thessaly to the south, was historically part of the Greek Kingdoms of Macedon and Thrace. However, it is essential to remember that during the time I am referencing, these kingdoms were not clearly defined, and people's perceptions of them varied. This understanding of ancient Greek geography might be one of the main reasons Orpheus is depicted in the earliest sources (starting in the 6th c. BCE) as a Pierian Greek and later as a Thracian Greek by ancient authors, depending on the era in which his name appears. For the ancient people of archaic Greece and later in Classical and Hellenistic times within the philosophical and religious spheres, Orpheus was a celebrated ancestor who lived in the Heroic Age. He was glorified for his pioneering advancements in music and revered as a theologian for his religious reforms. Commemorated for his role as a musician and priest among the Argonauts, he was immortalized for his heroic descent into the underworld to reclaim his wife, Eurydice, using the divine power of his voice.

Historically the Orphic tradition focused on themes concerning the human condition and the concept of an immortal soul. It emphasized the power of music, poetic language, ritual drama, and symbolism. It also explored the ideas of Remembrance and Fate and utilizes

https://doi.org/10.52202/078697-0002

invocation to foster a direct connection between the human soul, the natural world, and the divine. In this paper, I primarily aim to delve into the ancient idea known as "Theon Therapeian," which was one of the many therapeutic approaches in ancient Greece. Even though it is not exclusively an Orphic notion, it lies at the heart of both ancient and contemporary Greek religion and Mysteries. It emphasized "Service to the Gods," "Divine Worship," and the nurturing of the human soul through the worship of the Gods. Through this exploration, I will discuss the Orphic tradition in the context of contemporary Greek Mysteries and its relationship to Theon Therapeian. Additionally, I will discuss the transformative power of Divine Worship and the significance and application of ritual music in the Orphic tradition today. Finally, I will focus on specific uses of music in rituals and explore the philosophical connections between music and its role in facilitating the soul's ascent to divine realms within contemporary Orphic Mysteries.

The Orphic Tradition in Contemporary Times: A Brief Introduction, Key Beliefs, and Practices

The story of the Orphic tradition in contemporary times began at the start of the previous century, in the early 1900s. Two prominent figures, Spyridon Nagos (1868-1933) and Spyrodemos Anemogiannis Sinanidis (1933-2021), not only laid its foundations but also nurtured it through their followers and the writings they left behind. Spyridon Nagos, a renowned esoteric figure from Kakouri, Arcadia, played a pivotal role in reviving and structuring the Orphic and Pythagorean traditions in contemporary Greece (Νάγος, 2000, p. 9). He was a leader in esoteric teachings, authored influential works, and his approach to philosophy set high moral standards for all subsequent esotericism in Greece. Among his peers, he was admired for his ecumenical and humanistic approach, which is summarized in one of his sayings: "He who worships the Whole, also worships the part" (Νάγος, 2000, p. 31).

Conversely, Spyrodemos Anemogiannis-Sinanidis, an esteemed classical scholar and researcher from Athens, made significant contributions to the Orphic tradition ("Σπυροδήμος Ανεμογιάννης - Σινανίδης," n.d.). He taught philosophy at the National and Kapodistrian University of Athens, authored several books and academic papers on ancient Greek philosophy and subjects from the Orphic tradition, lectured extensively, and is regarded as the spiritual successor to S. Nagos. Through six generations of Orphics, their legacy persists today, extending beyond the confines of Greece. Within this single, undivided lineage of the followers of Orpheus, there are two primary approaches among the various Orphic Thiasoi and circles: the groups that

https://doi.org/10.52202/078697-0002

approach the Orphic tradition as a form of esotericism and those that view it primarily as an initiatic mystery religion. Both are undivided and interconnected among their adherents.

The Orphic tradition, rooted in ideas attributed to Orpheus in ancient Greek religious philosophies, diverged significantly from conventional Greek religious practices because of its unique doctrines and initiation rites. This divergence often attracted the scrutiny of the established religious authorities of the city-states. Such differences, notably the use of sacred texts during rituals and the abstention from meat, posed challenges. Although today's standards are radically different from those of ancient times, sacred texts such as the Orphic Hymns and the various Orphic Theogonies still occupy a central place in the rites and rituals of contemporary Orphics.

It is worth noting that when today's Orphics refer to a text or an idea as "Orphic," they are referencing the extant written materials traditionally attributed to Orpheus. Some of these include the contents of the Derveni Papyrus, the Orphic Hymns, the Orphic Fragments, and the Orphic Argonautica, to mention a few, as well as the oral Orphic traditions passed down within the Orphic lineage from generation to generation in Greece. This is regardless of contemporary academia's debate over whether these materials are genuinely Orphic. The writings of S. Nagos and S. A. Sinanidis are also included, and they are treated as sacred, with particular emphasis always placed on them. Key tenets of the Orphic tradition include a dualistic view of the soul and body, adherence to the cosmogonic myth of Dionysus Zagreus, and rigorous ritual practices. Among these practices is the approach known as Theon Therapeian, which will be explained later. It aims to achieve personal liberation, physical and spiritual purification, and direct communion with the God Dionysus, as well as with other Gods, through a complex combination of cultural practices that include ecstatic ritual music.

Continuing, it is essential to highlight the significance of the Orphic theogonies, particularly the *Orphic Hieros Logos in Twenty-Four Rhapsodies*. This rhapsodic poem, according to scholars, originates from the Hellenistic period to the 2nd century CE, but comprises materials and ideas that reference archaic and possibly even earlier times (Chrysanthou, 2020, p. 33-34). Presently, it survives only in fragmented form. At its core is the myth of Dionysus Zagreus, which recounts the creation of the universe, the generation of the Gods, the dismemberment of the young God by the Titans, their punishment by Zeus, mankind's birth from the remnants of the Titans and the Ether of Zeus, as well as Dionysus' rebirth from his father's thigh. This theogony

https://doi.org/10.52202/078697-0002

advances the notion that the human soul is immortal and contains a divine spark, but it is ensnared within a Titanic earthly body which the Orphics view as a tomb. This idea is also reiterated by Plato through the voice of Socrates in two of his dialogues, *Gorgias* and *Cratylus*. In a very characteristic and cyclical way, Socrates is echoing the Orphic notion of the body as the tomb of the soul: "Who knoweth if to live is to be dead, and to be dead, to live? And we really, it may be, are dead; in fact, I once heard one of our sages say that we are now dead, and the body is our tomb..." (Plato, trans. 1996, 492E).

Furthermore, it is essential to reference the cyclical arrangements of life, death, and rebirth which hold a central place in the Orphic Mysteries and religion. In the following Orphic fragment, the cyclical nature of the Orphic Dionysian eschatology becomes immediately apparent: "Life-death-life / truth / (a) Dionysus, Orphics" (Pajares & Cristóbal, 2008, p. 39). This conception drives the Orphic pursuit of purification to free the soul from the chains of reincarnation – a perception the Greeks termed metempsychosis and metensomatosis ("μετεμψύχωσις," n.d.). However, an important distinction must be made: The Titanic nature should not be confused with the notion of original sin. Both ancient and contemporary Orphics perceive the Dionysian and Titanic aspects as characteristic potential tendencies toward specific life modalities, not as inherent attributes from birth.

Another important practice is the custom of the ancient funerary texts found in the tombs of Orphic initiates. From various archaeological findings, it appears that ancient adherents often referred to short funerary texts, known today as the Orphic Gold and Bone Tablets. These were inscribed with petitions for acceptance by the Greek Gods of the underworld and instructions for the initiated in the afterlife, serving as guides. An example of such a tablet was found in an Orphic tomb in Pharsalus in the region of Thessaly in Greece. L 4 Tablet from Pharsalus (c. 350–330 B.C.E), National Museum of Athens:

> You will find in the mansion of Hades, on the right, a fountain
> and next to it, a white cypress erect.
> Don't get anywhere near that fountain!
> Further on you will Ànd, from the lake of Mnemosyne,
> water freshly flowing. On its banks there are guardians
> who will ask you what necessity brings you to them.
> And you will tell them absolutely all the truth;
> say: "I am the son of Earth and starry Heaven.
> My name is Asterius. I am dry with thirst; give me, then,
> to drink from the fountain." (Pajares & Cristóbal, 2008, p. 11).

https://doi.org/10.52202/078697-0002

This tradition continues today among Orphics who also incorporate them during the various Orphic rites and rituals performed regularly by the Orphic Thiasoi.

Finally, regarding the traditions of abstinence from killings and violence, it is said that Orpheus was the first to instigate such practices. As Aristophanes reiterates in his play *Frogs* through the words of Aeschylus: "Orpheus revealed mystic rites to us, and taught us to abstain from killings" (Aristophanes, trans. 2002, 1032). However, it is important to indicate that the ancient Orphics, Pythagoreans, and Platonists, who were at times indistinguishable from one another, were committed to a moral life of nonviolence and practiced vegetarianism and semi-vegetarianism.

Even though the degree to which such ways were practiced varied from person to person, it is well attested in ancient texts that they also had the piety and wisdom not to abstain from certain public and private animal sacrifices that were integral to societal harmony. The approach to this delicate matter is highlighted and reaffirmed by Porphyry of Tyre in his text *On Abstinence From Killing Animals* when he discusses Pythagoras and his followers in relationship to the subject of abstinence from meat: "And some say that the Pythagoreans themselves partake of animate creatures whenever they sacrifice to the gods" (*Porphyry*, trans. 2014, 104.26). Marinus of Neapolis indicates in his *Life of Proclus* that such was also the case with the philosopher and expert in Orphic matters, Proclus, who would permit himself to eat meat out of piety when sacrificing to the Gods and participating in certain cult settings (Marinus, trans. 1925, 19). It is important to note that while vegetarianism is also practiced among contemporary Orphics, it is left to the discretion of the individual's mode of living and their ability to adhere to such demanding standards of the tradition.

Today's Orphic rites and life practices are intended to nurture the divine element within, facilitating the soul's purification, eventual liberation from bodily constraints, and her reintegration with the divine.

"Theon Therapeian" as "Service to the Gods," and "Divine Worship" in Contemporary Orphic Mysteries

Prior to delving into the ancient idea termed Theon Therapeian and its relevance in the contemporary Orphic Mysteries, one must first contextualize the significance of therapy. This was a fundamental aspect of daily life for the ancient Greeks. In its broader sense, the Greek term

https://doi.org/10.52202/078697-0002

"θερᾰπ-εία", or "therapy" in English, was understood by the ancients to mean a form of "service" and "attendance". It had four principal applications:

1. Service and attendance of persons and service paid to the Gods,
2. As medical or surgical treatment or cure,
3. As care and tendance of animals, and
4. In a collective sense, a body of attendants or retinue. ("θεραπεία," n.d.)

Theon Therapeian, as an approach to therapy, can be found primarily in ancient Greek texts. It is expressed in various phraseological ways and attains a broad spectrum of meanings, some of which are still employed today in the same way as they were in ancient times. In its general sense, it is most properly understood as service paid to the Gods, conveying the development of an intimate bond with the Greek Gods through consistent ritualistic practices and worship. Socrates discusses this matter using his characteristic dialectic methodology in Plato's dialogue *Euthyphro*:

> S. All right, yet what kind of attendance on the gods would holiness be?
> E. It would be what slaves pay to their masters, Socrates.
> S. I see. It would be some kind of service to the gods, it seems.
> E. Very much so. (Plato, trans. 2017, 13d)

It is pertinent to note that in the wide-ranging context of Greek religion, the idea of "Service to the Gods" does not mean servitude the way we understand it today. The Greeks traditionally see their relationship with deity more as friendship rather than servitude in the strict sense of the word. The philosopher and priest of God Apollo at Delphi, Plutarch of Cheronia, mentions in his *Moralia: Advice to Bride and Groom* the above approach to deity in a very direct way: "The gods are the first and most important friends" (Plutarch, trans. 1928, 140.D.19).

Today, contemporary Orphic initiates have developed a deep and varied understanding of the idea of Theon Therapeian. This understanding encompasses a broad range of their religious and esoteric life, from establishing sacred places of worship to conducting votive processions. It also involves engaging in theurgy, performing rituals and dramas, composing music for ceremonial rites and contemplation, and executing sacred devotional acts. As I emphasized earlier, this is not part of a reconstructive effort but rather a meaningful evolution of a tradition that has been ceremoniously established, transmitted, and advanced, and is fully attuned to present-day needs.

https://doi.org/10.52202/078697-0002

In the following sections, I will delve into what is the essence of the contemporary Orphic Mysteries and esoterica in relationship to Theon Therapeian. Among the subjects to be discussed are the sacred acts of purification, ritual music, the use of invocations and hymns, and the use of ancient symbols.

Ritual Music in Contemporary Orphic Rituals and Practices

Music has always been at the heart of the Orphic tradition. With its origins deeply rooted in the life, times, and legend of Orpheus – who, according to tradition, charmed all of nature and even the Gods with his kithara – the Orphic tradition has consistently recognized the power of music to align the soul with the divine world. Orpheus's musical innovations are often mistakenly attributed to ancient Thracian shamanism and ancient Greek 'magic' by scholars like M.L. West (West, 1993, p. 4) and W.K.C. Guthrie (Guthrie, 1993, p. 39) due to cultural relativism. However, the true influence of Orpheus's musical innovations lies in intellectual, structural, and aesthetic advances they represent within the well-established ancient Greek musical tradition. The extraordinary abilities of Orpheus, as mentioned in ancient texts by authors such as Euripides, Aristophanes, and Apollonius Rhodius, are a poetic expression of admiration for his contributions to Greek music and culture. Rather than being contextualized with an emphasis on supposed "Thracian" shamanism and ancient Greek "magic," these abilities should be viewed as cultural and technical innovations. This is exemplified by his addition of the Hypate, the lowest string on the ancient lyre, attributed to him. Orpheus's innovations in the ancient melos, "melody, song," were also renowned in ancient times, so much so that they are referenced in Plutarch's *Moralia: On Music*, which highlights his distinctive and innovative musical style and treats him as a real person: "Terpander took as his models the hexameters of Homer and the music of Orpheus. But Orpheus evidently imitated no predecessor, as there were none as yet, unless it was composers of songs for the auloi, and Orpheus' work resembles theirs in no way" (Plutarch, trans. 1967, 1132F-1133).

At this point, it is important to note that "mousike," "music," in the context of Greek tradition and religion, encompasses not just the performance of instrumental compositions but also poetry, dance, and everything under the dominion of the Muses. This includes both the arts and sciences. Furthermore, music during the Orphic rituals does not merely fulfill an aesthetic role, but it signals the arrival of the Muses and the Gods related to music in the ritual. It is within this multidisciplinary approach to mousike, and through an elevated understanding of music as

https://doi.org/10.52202/078697-0002

having a divine origin, that contemporary Orphic practices reemerge. In the context of Theon Therapeian, these practices are consciously aligned with the Orphic, Dionysian, Pythagorean, and Platonic traditions, primarily manifesting as serious ceremonial and philosophical expressions, among other forms.

One of the most prominent applications of ritual music in contemporary Orphic rites is its use as medicine. Today, Orphic adherents believe that this notion, when properly understood and meaningfully approached, can have a direct therapeutic effect on the souls of participants during the rites. Historically, this idea is attributed to Pythagoras. Iamblichus, in his book *The Pythagorean Life*, transmits the following important passage regarding the therapeutic use of music by Pythagoras and his followers: "But at another time they used music in the place of medicine. And there are certain melodies devised as remedies against the passions of the soul, and also against despondency and lamentation, which Pythagoras invented as things that afford the greatest assistance in these maladies. And again, he employed other melodies against rage and anger, and against every aberration of the soul" (Iamblichus, trans. 1818, XX5).

Another central religious and esoteric approach to ritual music, widely used in today's Orphic rites, is the Harmony of the Spheres, an ancient Pythagorean idea that emphasizes celestial and musical order. Proclus, drawing inspiration from both Orphic teachings and Platonic philosophy, emphasized the cosmic significance of music. In his *Commentary on the Republic*, he describes harmony and music as tools in the hands of the demiurgic God Who uses them to create cosmic order as He fashions the Universe:

> He makes imitations both visible and beautiful of things both invisible and beautiful and makes imitations that exist naturally of the things that exist intellectually; using musical modes (harmonia) through which it is provided for the virtue that is in the whole to dominate and for vice to be diminished. He gives the rhythmic metre to the motions [in the universe] so that they undergo a motion that is rational (kata logon), and produces a single living harmony and one rhythm that is composed from all things. (Proclus, trans. 2018, 68.17 - 68.22)

The utilization of specific modes and rhythms, as described by Plato, Aristotle, and even later authors such as Proclus and Aristides Quintilianus, reflects various cultural attributes and moral states. Their main aim is to assist in intellectual and spiritual ascent and to shape the character of the soul in a transcendental manner. Additionally, the use of philosophical correspondences and analogies between numbers, music, and the sensible world plays an integral role in these practices. Essential rhythms employed during the Orphic rites include the

https://doi.org/10.52202/078697-0002

Paean, the Anapest, the Iamb, the Cretic, and the Bacchius. Moreover, the various astrological correspondences between the ancient Zodiacal signs and the associated musical notes, primarily derived from Platonism and Ptolemaic Astrology, are also considered significant. These elements form an important part of ritual music in the contemporary Orphic Mystery and esoteric traditions. Influential philosophers and authors in this domain, such as Pythagoras, Plato, Aristoxenus, Plutarch of Chaeronea, Proclus, Aristides Quintilianus, and Nicomachus of Gerasa, have continuously informed a vast range of contemporary Orphic musical practices. Later in this study, I will explore some of these methods, as I have personally contributed to the revival of these ideas, especially the Harmony of the Spheres and the purifications via the four classical elements in relation to ritual music.

One of the fundamental reasons for engaging in these practices today is the belief in the intimate connection between music, ethos, and the shaping of character and soul. The notion of music as an essential force in harmonizing the soul is evident in several important ancient texts, including Plato's *Republic*, where he states:

> "In that case, Glaucon," I said, "isn't an education in the arts most essential for these reasons, in that rhythm and melody above all penetrate to the innermost part of the soul and most powerfully affect it, bringing gracefulness, and, if one is brought up correctly, make one graceful; if not, isn't the result the opposite?" (Plato, trans. 2013, 401d)

This principle is applied to the core of Orphic practices, both ancient and contemporary.

In contemporary Orphic rituals, this ancient understanding is both revived and renewed. Ritual music is used to invoke specific deities, facilitate sacred acts, contemplative states, and create a sacred space for powerful devotional experiences. The choice of instruments, rhythm, melody, and even the silence between the notes is meticulously considered. This careful selection guides the practitioner on an immersive, transformative journey to achieve a state of ritual ecstasy and a sense of union with the Gods, known among the followers of the tradition as "enthousiasmos" ("Enthousiasmos," 2002). The ancient Greek lyre, the aulos, the frame drum, and the triangle are just a few of the traditional instruments used to achieve the inspiration and enthousiasmos sought by participants. Contemporary instruments, such as the guitar and synthesizers, are also used for their technological capabilities, though not as frequently as the traditional ones. Drawing from the wisdom of ancient musicians influenced by Orpheus, such as Terpander and Pindar, as well as polymaths like Pythagoras and Plato and the schools of thought they established, contemporary Orphics hold a unique perspective. They understand that music

https://doi.org/10.52202/078697-0002

is not merely an external sound, but a reflection of the inner and divine harmonies that could potentially affect a person's character and soul, and shape realities beyond that.

Four Mystery Rites Incorporating "Theon Therapeian" and Ritual Music – A Short Introduction

Below, I will attempt, to the best of my ability and within the limits allowed by tradition, to explore four contemporary Orphic Mystery rites. These rites are both religious and esoteric in nature and take place within the confines of sanctuaries and private homes. Present-day Orphics refer to their meeting places as "Orphic Sanctuaries" and "Ορφικές Οικείες" (Orphic Homes). The latter highlights the meaningful connection between the Orphic Mystery tradition and the structural dynamics of the Greek family unit. The following explorations aim to underscore the therapeutic, musical, and historical values of these rites and practices. I will also provide relevant philosophical insights accompanied by short musical excerpts to further deepen the understanding of the role of ritual music in the context of the Orphic rites.

The Purification Rite

In ancient Greek mystery traditions, as well as in contemporary mystery rites, the sacred act of purification occupies a central position in the lives of initiates and esoterics alike. Purification occurs not only during initiation rites but also in burial rites and everyday life, encompassing household worship. Within the framework of contemporary Orphic rites, purifications are commonly conducted during the early stages of initiation rites, irrespective of whether they are preliminary or represent a more advanced stage of the overall process.

The following brief description and analysis pertain to a preliminary purification rite that primarily occurs during the first stages of the initiation process. However, the formulaic nature of the rite encourages its repetition, hence it is employed at several junctures throughout the life of the initiate. This purification rite is primarily associated with the divine level of God Poseidon and is a poignant example of the contemporary Orphic Eleusinian elemental purification processes as a medium of the soul's transformation.

The initial part of the rite emphasizes the significance of water, an element often associated with rebirth and purification. As the initiand approaches the altar, they place both hands into a bowl filled with seawater to begin the first stage of purification through the cosmogonic element of Water. This immersion is an emblematic link to God Poseidon, who is usually referred to as the "second Zeus." Poseidon's dominion over the oceans and water bodies

 https://doi.org/10.52202/078697-0002

is not just an assertion of his power but represents the vast expanse of the element of Ether through the Cosmos. Furthermore, cosmogonic Water is one of the two primordial elements in the Orphic creation story and has been traditionally associated with purification since ancient times. The initiand's purification via the element of water suggests a cleansing of the deeper realms of the soul.

Following this sacred act, the pilgrim continues their journey on a designated Sacred Way until they meet the Priest and Priestess of Goddess Artemis and God Apollo for the next phase of purification. In the contemporary Orphic Mystery tradition, Goddess Artemis is associated with birth and rebirth but at a higher spiritual level, and God Apollo with enlightenment and the soul's freedom. The priests then perform an act of raising white candles, which symbolizes illumination, enlightenment, and the dispelling of darkness or ignorance. When they subsequently lower these candles to the heart position, it indicates an infusion of this divine light into the very core of the initiand's being. This act serves a dual purpose: it is a gesture of reverence and a means of purification by fire. The heart, often considered the seat of the soul and traditionally related to vitality and life force, is thus sanctified.

The rite concludes with another significant salutation and an act steeped in sacred symbolism. The initiand approaches the priest of God Zeus and steps on what is known as "Dios Kodion," a white sheepskin, with the left foot. The initiand acknowledges in reverence the protective embrace of the chief deity. The white fleece today has profound symbolism among contemporary Orphics, representing the vehicle of God Zeus's soul. It is also seen as a protective barrier, signifying purity at a divine level. This act also embodies the initiand's commitment to remain pure and grounded in divine wisdom. Finally, as the initiands, if more than one, depart, they carry with them the blessings and purifications of water, fire, and the divine, marking their spiritual ascent and alignment with the divine word.

The music used in the purification ritual varies; however, its primary objective remains consistent: symbolically purifying the pilgrims and instilling a sense of harmony, thereby preparing them for the forthcoming initiation processes and rites. Traditionally, during the purification rites, an ancient hymn known as the Paean is employed. This religious hymn is typically offered to Apollo, among other deities, who is invoked as a God of healing and purification. It is customary to offer the Paean at the beginning of a new undertaking and, as in antiquity, during other instances such as thanksgiving or to honor the God for granting victory

("Παιάν," n.d.). In these hymns, the Paeon which is the dominant foot of the meter comprises one long syllable followed by three short syllables (– ⌣ ⌣ ⌣) ("Paeon," n.d.). The preferred instruments are the ancient Greek lyre and the kithara primarily due to their association with Pythagoras, Orpheus and God Apollo.

The mode used in contemporary Paeans is the ancient Dorian mode (approx. E-E'in western notation), sacred primarily to God Apollo among other Gods. In the context of the Orphic tradition, the sacredness of this musical mode is mentioned in the Orphic Hymn 34 to the God Apollo, specifically between lines 19-23 (trans. 2013):

> ...now again with a Doric mode, harmoniously balancing the poles, you keep the living races distinct. You have infused harmony into the lot of all men, giving them an equal measure of winter and summer: the lowest notes you strike in the winter, the highest notes you make distinct in the summer, your mode is Doric for spring's lovely and blooming season.

Here is a brief example of music that incorporates a variation of the Paeanic rhythm in 5/4 and the ancient Dorian mode during the purification rites. The melody is composed for a seven-string lyre with Pythagorean tuning:

Ex. 1

"Myesis," "Initiation" – The Rite of the Enthronement

In the realm of ancient rituals, sacred rites from traditions deeply rooted in the ancient Orphic and Eleusinian Mysteries embody a myriad of symbolic representations, profound metaphors, and meticulous ceremonial procedures. During "Myesis," "Initiation," the first stage of the initiation process, the initiand approaches from the East, symbolizing new beginnings, enlightenment, and the divine realm. They are guided to a throne, usually a "Klismos," an ancient ceremonial chair associated with the Dionysian rites, or a "Skímpous," an Eleusinian ceremonial stool. This stool, historically favored by traveling pilgrims, symbolizes the transient and questing nature of those being initiated.

In the same ceremony, the initiand, undergoing a spiritual journey, has their head covered with a white "peplos" (veil) by the Mystagogue, their guide during the initiation process. This

https://doi.org/10.52202/078697-0002

white veil, indicative of purity and sanctity, provides a protective embrace while restricting the initiate's vision, symbolizing the transition from a world of darkness to the new light of the Mysteries. Priests and Priestesses, representing celestial forces, perform counterclockwise circumambulations around the enthroned pilgrim, a gesture symbolizing the rotation of the classical planets around the Sun and the idea of cosmic harmony.

Once enthroned, the initiand is presented with a torch by the Priest in their right hand. This torch serves as both a symbol of light in the darkness of the primordial night and a symbol of receptivity to divine wisdom. The Priestess then performs air and fire purifications, creating a sanctified atmosphere for the pilgrims. The final purification takes place when the High Priest places a thunderstone (black ophite) on the initiand's heart, signifying a deep spiritual connection with the divine mind and completing the purification process of the initiand's immortal soul. Interwoven in this rite is the Altar Priest's invocation of Rhea, Chthonic Demeter, and God Dionysos Zagreus, seeking blessings and protection on behalf of the initiand. Following these rites, the initiand, with vision restored, offers gratitude through tokens: an oil lamp to Rhea, wheat to Chthonic Demeter, and wine to Dionysos Zagreus. These offerings, emblematic of enlightenment, nourishment, and union with the Gods, echo the essence of the initiation process.

Concluding the rites, the newly made initiate, prepared for the second stage, which completes the process of initiation, the "Telete," is recognized as a transformed being and is gently guided back by the Mystagogue. This marks their successful journey into the sacred Mysteries. This combined rite serves as a liminal transition for the initiand from the realm of the uninitiated to that of the purified and the enlightened, surrounded by the presence of the Gods and guided by dedicated priests, priestesses, and other officials.

The "Myesis" or "Initiation" in the Orphic tradition, particularly the Rite of Enthronement, is a ceremonial process intimately connected with the Cretic and the Bacchius, ancient metrical feet used in ancient Greek poetry and dance during the ancient rites. There are several variations of these rhythms that can be performed at various speeds. A common form of the Cretic consists of three syllables: one long, one short, and one short at the end (– ˘ –) ("Κρητικός," n.d.). The Bacchius consists of three syllables: one short and two long (˘– –) ("Βάκχειος," n.d.). Today, there isn't an exact correspondence with Western music rhythm, but a close approximation to the Cretic and the Bacchius is the 5/8 rhythm, which is most used in the context of rituals today. The rite's music, following these rhythms, acts as an aural conduit to a sacred state, mirroring the initiate's transformation and spiritual ascent. Drawing upon the Dionysian ethos, as seen in Euripides' *Bacchae*, and deeply influenced by the ancient Corybantic tradition and dances, this rite employs music and movement as therapeutic expressions, reflecting the ecstatic nature of the Dionysian Mysteries and those of the Mother of the Gods, Rhea. In the Greek tradition, it is said that the Mother of the Gods used her ecstatic rites to cure God Dionysus from madness (see Apollodorus, trans. 1921, 3.5.1). Today, the Cretic and the Bacchius rhythms along with their variations are also used for their traditional therapeutic values against esoteric disharmony, emotional instability, and even depression. The ancient Phrygian mode (approximately D-D' in today's terms) was chosen for composing ritual music for the initiation because of its close association with the cult of Dionysos.

Below, you can find a brief illustration of how the Bacchius rhythm and the ancient Phrygian mode are employed in a contemporary, original composition during the initiation process:

Ex. 2

The Contemporary Asclepia Rites

Within the tapestry of Greek religion, the practices and beliefs associated with the Asclepius cult hold significant influence, resonating even in the contemporary era. The modern Orphic mystery rite, centered around the ancient Asclepius and his family, serves as an epitome of renewal and wholeness, harking back to an age of divine intervention in health and wellbeing. Through structured phases, such as the Pombe and subsequent invocations, this ritual seeks to encapsulate the healing essence and powers of the deity, calling upon the benevolence of the God of Healing, Asclepius, and his family.

Embarking on the path of purification, the ritual begins with a devotional "Pombe," (procession) an age-old custom that prepares participants by distancing them from the mundane and sanctifying them for their encounter with God Asclepius during the rites.

Following the Pombe, the next step involves the purification of the sacred space, in order to establish the place as appropriate for ritual. This process invokes the purity and serenity that

https://doi.org/10.52202/078697-0002

the ancient Asclepia sanctuaries were known for, fostering an environment that encourages the restoration and ascent of the soul.

As participants enter this sacred boundary, the ritual progresses to the opening invocations, specifically addressing Hygeia, the Goddess of health, cleanliness, and hygiene, Asclepius, the savior of health and healing, and God Apollo, who in the Orphic tradition is acknowledged as the father of Asclepius. These initial invocations have great significance since their function is to assist the High Priest in petitioning for the attendance of the Gods during the rites. At this juncture, the recitation of appropriate Orphic hymns occurs. These hymns, ancient and profoundly deep in meaning, create a resonance that connects the pilgrims with the divine, fostering a reciprocal communication and communion with the help of divine Eros. The recitation of the hymns is followed by the placement of sacred objects, notably a reverent installation of Asclepius' cult statue, an act that serves to physically manifest the deity's presence amongst the participants, known as Epiphany.

As the ritual further unfolds, personal purifications are undertaken. Participants engage in the symbolic act of cleansing their hands and face, echoing ancient customs that underscore the purification of the body as the vessel of the soul. Following this, the practice of "Enkoimisis," a form of incubation primarily understood as "ritual sleep," takes precedence. This phase represents the act of lying down in the sacred spaces of Asclepius, a profound gesture that symbolizes surrender and readiness to receive the deity's healing. Attendants typically participate in this stage while sitting, immersing themselves in a deep contemplative state. Contemporary adherents believe that this practice paves the way for participants to fully immerse themselves in the deepest recesses of their souls. As the ritual culminates, the pilgrims engage in sacred acts of worship, acknowledging and venerating the divine presence, solidifying their connection and communion with the Gods. In a final act of devotion, offerings are made at the altar, a symbolic gesture of gratitude and a plea for grace and healing. Finally, during these sacred rites, the therapeutic value of music is meaningfully integrated and considered, not only on behalf of the "Orpheotelestai" or "Orphic ritualists," but also by all those who are attending the rites.

In today's expression of Asclepia, music primarily conveys a slow ceremonial tone. While a variety of rhythms and musical landscapes are used in these rites, the Paean, the ancient Dorian mode, the ancient Paeon Epibatos rhythm, and the ancient Greek lyre, as observed in purification rituals, hold prominent positions. The Paeon Epibatos foot is a metric rhythm that consists of five long syllables (– – – – –), roughly corresponding to the 5/4 time signature in Western music notation (West, 1992, p. 156). However, the ancient concept of purification through the four classical cosmogonic elements is also a significant part of ancient Greek hieratic art. Therefore, it is incorporated into the music composed for the Asclepia, running in parallel with the sacred purification acts that occur during the rites. Proclus, in his work *On the Timaeus of Plato: Book 1,* while providing his own exegetics, openly emphasizes this ancient hieratic practice:

> The one [type of] purgation arises through water, the other through fire, and in every case, purity comes to the second things thanks to the first. Hence, in Orpheus' work too, Zeus is required to bring the purgatives from Crete, as the theologians usually interpret Crete as representing the intelligible (Proclus, trans. 2007, 118.21 – 118.26).

I first introduced the above-mentioned approach to music via original musical compositions into the Orphic tradition a few years ago, and it seemed to have a great appeal and a positive effect on the purificatory experience among contemporary Orphics. For that reason, it was integrated into the tradition in a very short amount of time. However, it is important to note that even though the way the subject is approached musically today is original, the idea itself is not; it has its roots in Orphic, Pythagorean, and Platonic philosophy and practices. To integrate the idea of purification via the elements into the music composed for the rites without affecting the ancient principles that govern the most important elements of the composition, such as the choice of scales and rhythms, philosophical correspondences between certain musical notes or series of notes, and the cosmogonic elements are carefully employed to establish the appropriate elemental hierarchical order and the various analogies between the elements. For example, to musically highlight the idea of purification through water, one can incorporate the tetrachord meson into the opening section of a musical composition for the lyre: [B, c, d, e]. Later in the composition, the tetrachord diezeugmenon can be used to emphasize purification by the element of fire: [b, c1, d1, e1]. The tetrachords mentioned are part of the Greater Perfect System of ancient Greek music (Michailidis, 2003, p. 231). The philosophical correspondences between musical tones and the cosmogonic elements, as mentioned above, are preserved in Aristides Quintilianus' book *De Musica, Book III*:

> We also find that the five primary elements are analogous to the tetrachords, earth, being heaviest *[barytaton],* to the tetrachord *hypaton,* water to the tetrachord *meson,* since it is closest to earth, air to the tetrachord *synemmenon.* (For air sinks down and is drawn down into the depths of the sea and the caves of the earth to enable creatures there to

https://doi.org/10.52202/078697-0002

breathe, since it is essential to them, and without it they would not survive.) Fire is analogous to the tetrachord *diezeugmenon,* since movement downwards is contrary to its nature and it can be drawn downwards only by an external force, while its natural movement is upwards, away from things on the earth. The aether should be assigned to the
tetrachord *hyperbolaion,* since aether is the highest. (Barker et al., 1984, p. 515).

Below is a short musical passage employing a slow Paeon Epibatos and the elements:

Ex. 3

The Orphic Platonic Harmony of the Spheres Rite

The Orphic Platonic Harmony of the Spheres Rite is an esoteric initiation, constituting the second stage of a three-stage initiation process. It pays homage to the ancient Orphic-Pythagorean and Platonic cosmology, philosophy, and theology, particularly as expounded in the tenth book of the Republic, within the myth of Er. This rite, deeply rooted in Ancient Greek religious customs, primarily embodies Orphic liturgical traditions. Its purpose is to instill a sense of internal and cosmic harmony in the pilgrim's mind and their interrelation.

At the heart of the rite lies a sacred altar, venerating several ancient Greek deities, including the Olympians associated with the seven classical planets from Plato's time. The rite commences by establishing a sanctified space, participant purification, altar purification, and an opening invocation to the cosmic principle of Number Eight, reflecting Pythagorean thought. In contemporary Orphic tradition, this number symbolizes the Law of Harmony in the Cosmos and Goddess Aphrodite, perceived as the mother of Goddess Harmonia and the divine consort of God Ares.

Participants then recite Orphic Hymns, invoking the benevolence of the Gods. Each hymn serves as an epiclesis, symbolically drawing the pilgrims closer to the divine essence, allowing unity between the participant and divinity through intellect. The core of the rite encompasses sacred acts and dramatic illustrations, commencing at the Gate of the Moon, associated with Cancer in the Zodiac, and culminating at the Gate of Saturn in Capricorn. The journey of the initiate mirrors the ascent of the Platonic soul through the seven celestial spheres. Each sphere,

https://doi.org/10.52202/078697-0002

represented by a candle on the altar and a Priest or Priestess, embodies specific cosmological principles and their related deities. The Gate of Saturn, symbolizing the outermost realm of the Cosmos, leads to the Empyrean, where Platonic Forms reside. Here, the soul draws near to the divine levels, encountering Goddess Harmonia. Her symbol, two intertwined snakes, represents cosmic union and balance between the two principal cosmogonic substances, Water and Earth, signifying the alignment of the initiate's soul with the Divine Mind and the World Soul, resonating with cosmic harmony. The rite concludes with the High Priest lighting a torch (or candle), symbolizing the fulfillment of cosmic order at the divine levels.

The ritual music, reflecting the doctrines of Pythagorean Harmony of the Spheres, creates a harmonic microcosm of the universe in the ritual space through the laws of similarity, analogy, and the art of correspondences. This alignment guides the initiates to contemplate the idea that the principle of harmony is essential for the constituent parts of the World Soul to coexist in unity. Similarly, and by extension, the human soul, with its celestial origin, requires harmony to unify its elements, representing a microcosm of the World Soul.

Finally, the ancient Greek notion that musical scales and rhythms influence a person's ethos and character is also considered. While it is beyond the scope of this study to provide a detailed exploration of the aforementioned idea, I will share a few notable ancient references below for clarity and to assist the reader in relating to them more effectively. This will be accompanied by a brief description of the compositional technique most used for the music in these rites. Regarding the significant role of harmony as a unifying principle in the context of ancient Greek cosmology, one of the fragments attributed to the Pythagorean Philolaus relates the following: "But since the first principles were not by nature alike or akin, it would be impossible for them ever to have been arranged, had not harmony supervened, in whatever way it came into being" (Kirk et al., 1983, p. 310). Also, on fragment 18.22 the soul is equated to a form of harmony: "Philolaus <said that the soul> is a *harmonia*" (McKirahan, 2011, p. 363).

In ancient Pythagorean cosmology and Platonic thought, the planetary spheres mentioned in the description of the rite above are considered important parts of the cosmic soul. Their coordination by divine harmony makes the idea equally important for Orphic initiates to experience and participate in because it is believed to be familiar to the soul due to its divine origin. For this reason, the ancient eight-note scale (approximately E-E'), which is attributed to Pythagoras and the planets associated with the notes in that scale, are used. They highlight the

https://doi.org/10.52202/078697-0002

planetary positions along with the sphere of the Zodiac as they move through the night sky live during the initiation process. The ancient eight-note Dorian mode is the musical mode most used in this rite due to its association with Pythagoras, God Apollo, and the cosmic harmony He provides through it. The ethos of that scale is also relevant, as it is said to inspire a feeling of splendor, nobility, and courage. The philosopher Plutarch, while discussing the modes in his work *Moralia: On Music*, makes two important revelations regarding their ethos. The first revelation pertains to the significance of musical expertise and knowing how to use the musical scales: "For one who knows the Dorian mode without the skill to pass judgement on whether it belongs here or there will not know what effect he is producing; in fact, he will not even preserve the moral character of the mode" (Plutarch, trans. 1967, 1143c). The second concerns Plato's preference for the Dorian mode: "...for Plato had studied the science of music with great care, receiving instruction from Dracon of Athens and Metellus of Agrigentum. No, he preferred the Dorian because, as I said earlier, it has a preponderance of noble gravity" (Plutarch, trans. 1967, 1136). The rhythm of the compositions for this is usually slow and ceremonial, and for that reason, the ancient Anapestic rhythm is selected. Today, this rhythm is primarily used in ceremonies and slow processions.

In its most basic form, the approach to creating Harmony of the Spheres music is primarily melodic, involving the following steps: The melody is simple, constructed from a series of notes that correspond to the planets as they rise, reach their respective midpoints in the heavens, and then set. In the example below, each sequence adheres to the order based on the respective timings of the rise, meridian, and set of each celestial body on the 14th and 15th of May 2024, as observed from Delphi, Greece. This perspective is from an Earth-based observer. The selected date roughly aligns with the lunar day marking the 'birth' of the God Apollo, the seventh day of the Attic month Thargelion, according to the sacred Greek Attic calendar as it is practiced today. The preferred rhythm for Harmony of the Spheres music is a slow Anapestic rhythm. An Anapest is a metrical foot comprising two short syllables followed by one long syllable (⏑⏑–) ("Anapest," 2016) and in its most common form, it corresponds to today's 2/2 rhythm (West, 1992, p. 136).

https://doi.org/10.52202/078697-0002

The Scale/Planets Relationships for the Ancient Pythagorean Dorian Octachord

(see Nicomachus, trans. 1994, Ch. 5)

1. Fixed Stars - E (High E')
2. Saturn - D
3. Jupiter - C
4. Mars - B
5. Mercury - A
6. Venus - G
7. Sun - F
8. Moon - E (Low E)

The Planets on May 14/15, 2024

Location: Delphi, Greece.

Rise (May 14, 2024):

1. Mars - B (Rises at 4:20 am)
2. Mercury - A (Rises at 5:15 am)
3. Venus - G (Rises at 6:01 am)
4. Sun - F (Rises at 6:13 am)
5. Jupiter - C (Rises at 6:29 am)
6. Moon - E (Low E) (Rises at 11:41 am)
7. Saturn - D (Rises at 3:27 am on May 15)

Mid-heaven (Meridian) (May 14 and 15, 2024):

1. Mars - B (Meridian at 10:34 am on May 14)
2. Mercury - A (Meridian at 11:44 am on May 14)
3. Venus - G (Meridian at 12:58 pm on May 14)
4. Sun - F (Meridian at 1:20 pm on May 14)
5. Jupiter - C (Meridian at 1:34 pm on May 14)
6. Saturn - D (Meridian at 9:08 am on May 15)
7. Moon - E (Low E) (Meridian at 7:09 pm on May 14)

Set (May 14 and 15, 2024):

1. Mars - B (Sets at 4:47 pm on May 14)
2. Mercury - A (Sets at 6:13 pm on May 14)
3. Venus - G (Sets at 7:57 pm on May 14)
4. Sun - F (Sets at 8:29 pm on May 14)
5. Jupiter - C (Sets at 8:40 pm on May 14)
6. Moon - E (Low E) (Sets at 2:27 am on May 15)
7. Saturn - D (Sets at 2:49 pm on May 15)

("Planets Visible in the Night Sky in Delphi, Central Greece, Greece," n.d.)

https://doi.org/10.52202/078697-0002

Here is an example of a composition that incorporates the planetary associations and conditions for May 14/15, 2024, along with a variation of a slow ceremonial, ancient Anapestic rhythm. The lyre has been tuned in to the Dorian mode with a Pythagorean tuning:

Ex. 4

Final Remarks

In this examination of the role of the Orphic tradition in contemporary Greek Mysteries and esotericism, I have explored the enduring relevance and transformative power of ancient Greek thought and Orphic practices in contemporary Greek religion and spirituality. Central to my study was the ancient idea of "Theon Therapeian," an ancient therapeutic approach rooted in the service and worship of the Gods, encapsulating the essence of the Orphic way of life aimed at achieving inner harmony and divine union.

The revival of these ancient Mysteries and the esotericism produced by figures like Spyridon Nagos and Spyrodemos Anemogiannis-Sinanidis illustrates a strong yearning for spiritual depth and a connection with Greek ancestral wisdom and cultural practices. This resurgence is not merely a nostalgic return but a vibrant, evolving tradition with coherent continuity that integrates ancient insights with current spiritual needs.

Particularly important is the role of ritual music in the Orphic tradition. Far from being a mere cultural relic, it emerges as a potent agent for healing and transformation, harmonizing the human soul within and with the divine in the context of ritual. Furthermore, through this brief but meaningful exploration of the ancient and contemporary Orphic tradition, my aim is to assist in revitalizing interest in the immense therapeutic value and range of its practices. In my opinion,

https://doi.org/10.52202/078697-0002

having experienced them firsthand, they promote a more fulfilling life, not only on an individual but also on a communal level. As we navigate our contemporary world, the insights drawn from the contemporary Orphic way of life offer a rich source of inspiration and wisdom, with the potential to guide us toward a more profound and harmonious existence.

https://doi.org/10.52202/078697-0002

References

Anapest. (2016, July 20). In The Editors of Encyclopedia Britannica (Ed.), *Encyclopedia Britannica*. https://www.britannica.com/art/anapest (accessed Dec. 20, 2023)

Barker, A., Stevens, J., & Huray, P. L. (1984). *Greek musical writings: Volume 2, harmonic and acoustic theory*. Cambridge University Press.

Chrysanthou, A. (2020). *Defining Orphism: The beliefs, the ›teletae‹ and the writings*. Walter de Gruyter GmbH & Co KG.

Enthousiasmos. (2002). In H. Cancik, H. Schneider, & C. F. Salazar (Eds.), *Brill's New Pauly*. http://dx.doi.org/10.1163/1574-9347_bnp_e12221560

Guthrie, W. K. (1993). *Orpheus and Greek religion: A study of the orphic movement*. Princeton University Press.

Kirk, G. S., Regius Professor of Greek G S Kirk; F.B.A., Raven, J. E., & Schofield, M. (1983). *The Presocratic philosophers: A critical history with a selection of texts*. Cambridge University Press.

McKirahan, R. D. (2011). *Philosophy before Socrates: An introduction with texts and commentary*. Hackett Publishing.

Michailidis, S. (2003). *Encyclopedia of Ancient Greek Music*. National Bank of Greece Cultural Foundation.

Paeon. (n.d.). In *The Online Liddell-Scott-Jones Greek-English Lexicon in Thesaurus Linguae Graecae® Digital Library*. University of California, Irvine. https://stephanus.tlg.uci.edu/lsj/#eid=78951 (accessed Dec. 17, 2023)

Pajares, A. B., & Cristóbal, A. I. (2008). *Instructions for the netherworld: The orphic gold tablets*. BRILL.

Planets Visible in the Night Sky in Delphi, Central Greece, Greece. (n.d.). timeanddate.com. https://www.timeanddate.com/astronomy/night/@11287748 (accessed Dec. 20, 2023)

Βάκχειος. (n.d.). In *The Online Liddell-Scott-Jones Greek-English Lexicon in Thesaurus Linguae Graecae® Digital Library*. University of California, Irvine. https://stephanus.tlg.uci.edu/lsj/#eid=19628 (accessed Dec. 17, 2023)

West, M. L. (1992). *Ancient Greek music*. Oxford University Press.

Σπυροδήμος Ανεμογιάννης - Σινανίδης. (n.d.). Academia.edu. https://independent.academia.edu/ΣπυροδήμοςΑνεμογιάννηςΣινανίδης (accessed Dec. 15, 2023)

θεραπεία. (n.d.). In *The Online Liddell-Scott-Jones Greek-English Lexicon in Thesaurus Linguae Graecae® Digital Library*. University of California, Irvine. https://stephanus.tlg.uci.edu/lsj/#eid=49703 (accessed Dec. 15, 2023).

Κρητικός. (n.d.). In *The Online Liddell-Scott-Jones Greek-English Lexicon in Thesaurus Linguae Graecae® Digital Library*. University of California, Irvine. https://stephanus.tlg.uci.edu/lsj/#eid=61806 (accessed Dec. 17, 2023)

μετεμψύχωσις. (n.d.). In *The Online Liddell-Scott-Jones Greek-English Lexicon in Thesaurus Linguae Graecae® Digital Library*. University of California, Irvine. https://stephanus.tlg.uci.edu/lsj/#eid=69291 (accessed Dec. 15, 2023)

Νάγος, Σ. Κ. (2000). *Ο κάτοικος του σύμπαντος*. Ιδεοθέατρον.

Παιάν. (n.d.). Logeion, University of Chicago. https://logeion.uchicago.edu/Παιάν (accessed Dec. 17, 2023)

 https://doi.org/10.52202/078697-0002

Deep Breathing as a Tool to Promote and Enhance Flow State

Adina Dabija

Abstract

This research was conducted to investigate if voluntary regulation of breath can facilitate entering or enhancing flow state. Deep breathing, also known as yogic or diaphragmatic breathing, consists in consciously and actively using the diaphragm to increase the inflow and outflow of air, as well as to decrease the frequency of each breathing cycle (Russo et al., 2017), as a potential flow state promoting or inhibiting tool via vagus nerve stimulation and other physiological mechanisms. We hypothesized that, since deep breathing is an efficient way to promote relaxation, oxygenate the body, and reduce stress, it can be used as a simple, free, and actionable tool to create an optimal mental state when the sympathetic activity is too intense to enter or maintain flow (i.e. by a challenge perceived as superior to the skill), as well as a post-flow aftercare technique to limit the negative effects of the physiological arousal state, such as increased sympathetic nervous system arousal, which could potentially lead to stress-related cardiovascular effects (Murch et al., 2020) and high cortisol (a stress hormone) levels (Peifer et al., 2015).

Keywords: Flow State, Deep Breathing, Peak Experience

https://doi.org/10.52202/078697-0003

Introduction

The term "flow state" was coined by the psychologist Csíkszentmihályi in his groundbreaking book *Beyond Boredom and Anxiety* (Csíkszentmihályi, 1975). The author defined flow as "a state in which people are so involved in an activity that nothing else seems to matter; the experience is so enjoyable that people will continue to do it even at great cost, for the sheer sake of doing it" (p. 71). Flow state is a pleasant state of absorption of a person during an optimally challenging activity. During flow, time perception is typically distorted, and self-referential thoughts fade out of the mind as a subject shows undivided attention to a limited stimulus field. This experiential state occurs when skills of a person and demands of the activity are in balance, and both above average (Csikszentmihalyi & LeFevre, 1989; Rheinberg, 2008).

Csíkszentmihályi, a distinguished Professor of Psychology and Management at Claremont Graduate University, previously holding the position of chief of the department of psychology at the University of Chicago, was one of the world's leading researchers in positive psychology, a branch of the field focused on optimizing behaviors for success. He wanted to investigate flow state as a possible antidote to the post-World War II existential anguish that many experienced at that time, as most people dealt with loss of family members, jobs, homes, and the overall sense of safety and purpose. He developed a questionnaire on activities that produce the deepest enjoyment and the greatest satisfaction and, after analyzing the data, he concluded that when people are highly concentrated and optimally challenged while being in control of the action has the most positive impact even when the tasks are lacking external rewards. Csikszentmihalyi and LeFevre (1989) called the flow experience the "optimal experience" and later mentioned happiness as part of flow. This idea was also reinforced in a later article, which the authors stated: "Flow is defined as a psychological state in which the person feels simultaneously cognitively efficient, motivated, and happy" (Moneta & Csikszentmihalyi, 1996) (p. 277). This is relevant because, according to Csikszentmihalyi, most people spend more than half their time in a dissatisfied state of mild anxiety and mind wandering, particularly during their unstructured leisure time. "Structuring time around building a sense of autonomy, mastery and purpose in work and leisure activities lead to the most significant increase in overall happiness" (Csikszentmihalyi, 1975).

https://doi.org/10.52202/078697-0003

A Historical Perspective on Flow

Prior to Csíkszentmihályi, Maslow (1954) investigated the concept of "peak experience," described as "a moment of awe, ecstasy, or sudden insight into life as a powerful unity transcending space, time, and the self" (p. 160). In his book *Motivation and Personality*, Maslow studied exemplary "self-actualized" people and identified some of their common traits. One of them is the ability to have "peak experiences," such as Albert Einstein sailing to the middle of Lake Geneva to gaze at the stars, in wonder. Maslow proposed that this type of experience helped Einstein "crack the code" of the universe. He found that peak experiences were more likely to occur for individuals in their pursuit of self-actualization, their internal drive and the hyperfocus that come with it representing its own reward and less likely in individuals who are unmotivated and non-striving. Following his research on self-actualized individuals, Maslow coined the term "positive psychology" in *Motivation and Personality*, in contrast with the psychological approaches focused on pathologies. Maslow believed that psychology's preoccupation with disorder and dysfunction lacked an accurate understanding of the human potential and he emphasized the importance of studying multiple aspects of an individual, especially those related to their potential for growth.

Flow was probably achieved in shamanic trace or other types of mystic ecstatic practices across millennia. Ancient Greeks believed that being divinely inspired was foundational for any act of creation. The seven muses, ruled by Apollo, the god of medicine, music and divine guidance, were invoked before artistic endeavors in order to achieve divine inspiration - θεόπνευστος - theópneustos (Theos - God and pneuma - dynamic movement of air or breath, literally meaning "God-breathed" or "divinely inspired"). It is worth mentioning that the word "music" derives from the Greek word "moῦσαι" - muses - which were thought to be goddesses who preside over the process of creation in arts and sciences, bringing humans remembrance of joy and forgetfulness of pain.

Dao De Jing, a sacred text from Ancient China, attributed to Lao Tzu, introduced the concept of "wu wei" 無爲, literally meaning "inexertion," "inaction," or "effortless action," being at one with an activity performed in a state of profound plasticity: "That which offers no resistance, overcomes the hardest substances/. Those that have no substance can penetrate the

https://doi.org/10.52202/078697-0003

solid./ Therefore I know non-action wins success./ Teaching without words,/ Succeeding without action,/ These are understood by the very few" (Lao Tzu, 2020, ch. 43).

Flow was probably experienced by Sufi mystics when they entered "hal" - a state of ecstasy in which the divine became manifest. During the Islamic golden age, when the Islamic world leaped ahead in mathematics, poetry, and the sciences, building on a foundation of Greek and other ancient texts translated into Arabic, many innovators freed themselves from authorship to "merge with the divine" through their art and inventions (Frishkopf, 2003). With the advent of the Renaissance, arts and sciences started to flourish in Europe and extraordinarily gifted individuals in flow such as Michelangelo, Leonardo Da Vinci, or Giordano Bruno helped shift Western values towards a freer and more open-minded view of the world in an ultra-conservative environment strictly controlled by the church.

Method

For the first part of the study, we reviewed literature that described psychological and neurobiological models of flow. Our search was not intended to be exhaustive, but included the most recent studies of Google Scholar, Web of Science, and PubMed. Key words used to retrieve relevant studies included "characteristics of flow," "neurobiology of flow," and "flow and sympathetic nervous system," and "flow and vagus nerve."

For the second part of the study, we reviewed the literature on vagus nerve and breathing and breathing and flow. Then we conceptualized a model of voluntary control of breath to optimize flow based on the neurobiological models of flow found in the first part of the study. Our hypothesis is that deep breathing could be used as a simple and efficient way to promote relaxation when the sympathetic activity is too high to enter or maintain flow (i.e. by a challenge perceived as superior to the skill), as well as in the post-flow aftercare, to limit the negative effects of the physiological arousal state, such as increased sympathetic nervous system arousal, which could potentially lead to stress-related cardiovascular effects (Murch et al., 2020) and high cortisol (a stress hormone) levels (Peifer et al., 2015). We hope this hypothesis will be tested in a future study.

Results

In the original flow model, described by Csikszentmihalyi in 1975, flow occurs in the goldilocks area of the balanced challenge/skill ratio (Fig. 1, left-hand): in this state of balance, one feels both

https://doi.org/10.52202/078697-0003

optimally challenged and confident that the task to fulfill is under control. When a challenge is too low for the skill, boredom arises; when the challenge surpasses the skill, anxiety arises. Due to theoretically inconsistent results, this model was reformulated by Csikszentmihalyi and Csikszentmihalyi (1988). The revised model (Fig. 1, right-hand) proposes that flow is experienced only when challenge and skill are both high. This model is sometimes referred to as the "four channel model," or quadrant model.

Fig. 1 Original Flow Model (left-hand side; Csikszentmihalyi 1975) and reformulated quadrant Model of Flow (right-hand side; Csikszentmihalyi and Csikszentmihalyi 1988)

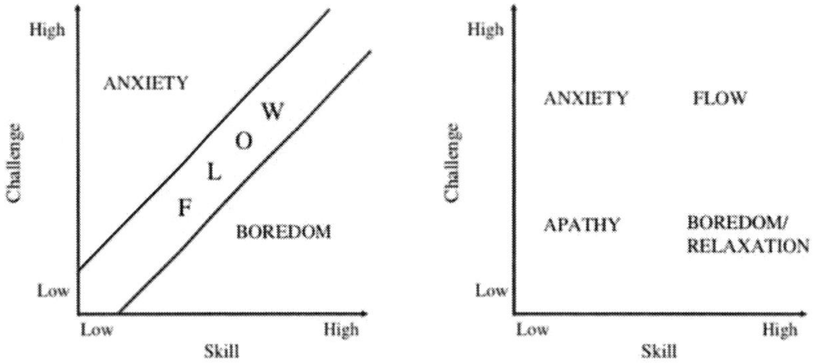

However, even this model presents conceptual challenges. First of all, not everyone has the same propensity to experiencing flow states. Individuals intrinsically motivated such as creative individuals and high achievers have a higher propensity to flow, described by Csikszentmihalyi (1975, 1990) as having an autotelic personality. On the other hand, neurotic personality types, those with high self-criticism and those experiencing high levels of stress and anxiety have a lower one (Ullén et al., 2012). Empirical evidence reported by Moneta and Csikszentmihalyi (1996) also pointed to other individual differences. They found that the balance of challenge and skill does not go hand in hand with high values in the flow indicators such as high concentration for all individuals. When looking at the achievement motivation research, the individual differences could easily be explained by differences in the achievement motive: the assumption that some people experience balance as positive and some as negative forms the core of the risk-taking model of Atkinson (1957); Brunstein (2008). According to this model, highly achievement-motivated individuals prefer tasks of medium difficulty (e.g. tasks in which the balance of difficulty and skill is present). In contrast to this hope of success-aspect of the achievement motive, individuals with a strong motive of fear of failure even avoid tasks of medium difficulty. The assumption that the achievement motive moderates the effects of the balance seems even more plausible considering that "the flow model may be more applicable to social contexts and activities where achievement plays a dominant role" (Moneta & Csikszentmihalyi, 1996).

Recent researchers started to study flow not only based on the one component (the skill-challenge ratio), evaluated based on the questionnaires in the Appendix, but also on physiological indicators such as heart rate variability and cortisol levels. From a neurobiological perspective, flow is a state of relaxed but heightened arousal marked by a "yin-yang" balance of the two branches of the autonomic nervous system - the sympathetic nervous system with its "fight-or-flight" mechanism and the parasympathetic nervous system, with its "rest-and-digest" mechanisms. Anxiety and stress are states of high physiological arousal, as indicated by increased sympathetic nervous system and HPA-axis activation; boredom and relaxation are characterized by low physiological arousal.

Fig. 2 The relationship of flow-experience and physiological arousal is an inverted u-curve (Peifer, 2014)

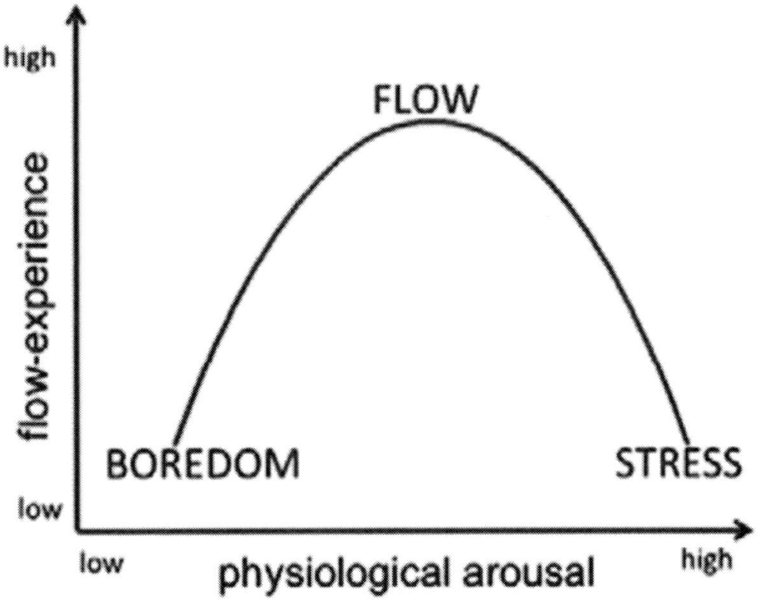

https://doi.org/10.52202/078697-0003

Peifer et al. (2014) found that the highest flow values occurred when high parasympathetic activation was combined with moderate sympathetic activation, which supports their hypothesis of an inverted u-shape relationship between flow and sympathetic activation, where more physiological arousal is associated with reduced flow. These results contradict findings of Keller and colleagues, who found a negative relationship between flow and parasympathetic activation (Keller et al., 2011). Peifer et al. concluded that flow is associated with moderate forms of arousal that are experienced as challenging and it is reminiscent of moderate and positive forms of stress, known as "eustress."

A study on transcutaneous stimulation of the vagus nerve during flow showed that direct stimulation of the vagus nerve during flow state resulted in inhibitory effects on maintaining flow (Colzato et al., 2018). This undoubtedly shows that the vagus nerve is modulating flow. The vagus nerve is part of the parasympathetic nervous system, which is responsible for a vast array of crucial bodily functions, including control of mood and stress response. Thus, this study provides evidence for a positive relation between flow and parasympathetic activation.

One key physiological mechanism of the flow is the locus coeruleus - norepinephrine system, which regulates task engagement through different modes of norepinephrine release. This system is crucial for maintaining the balance between task engagement and disengagement, with flow states being associated with intermediate levels of arousal, fitting an inverted U-shaped relationship.

Additionally, electroencephalogram (EEG) studies have shown that flow states are associated with increased frontal theta activity and moderate frontocentral alpha rhythms. These EEG patterns suggest high levels of cognitive control and immersion in the task, without excessive working memory load (Katahira et al., 2018).

The interaction of large-scale brain networks also plays a significant role. The Default Mode Network (DMN), Central Executive Network (CEN), and Salience Network (SN) are involved in the strong task engagement and low self-referential thinking typical of flow. Dopaminergic and noradrenergic systems mediate the intrinsic motivation and positive mood states associated with flow (Van der Linden et al., 2021).

Discussion

Strategic deep breathing during the flow activity at intervals when the actor feels a decrease in performance could relieve some of the stress and anxiety that hinders entering and maintaining

https://doi.org/10.52202/078697-0003

flow. Vagal activation, resulting in an increase in dopamine levels in the brain, leading to a greater sense of pleasure, along with overall increase in body oxygenation, would support the performer's effort to meet the challenge perceived as overpassing the skill. Deep breathing has been shown to influence various physiological parameters that are closely linked to the flow state. For instance, deep breathing can modulate cerebral circulation and enhance cerebral blood flow, as demonstrated by Li (Li et al., 2023). This improved cerebral circulation can facilitate cognitive functions and mental clarity, which are essential components of the flow state.

Additionally, deep breathing has been found to affect the autonomic nervous system by promoting parasympathetic activity and reducing sympathetic activity. This autonomic modulation can lead to physiological relaxation and improved heart rate variability, as discussed by Noble and Hochman (2019). Enhanced heart rate variability is associated with better emotional regulation and cognitive performance, both of which are critical for achieving a flow state.

Deep breathing has been proven to reduce stress in adults (Hopper et al., 2019). According to a January 2023 study published in Cell Reports Medicine (Balban et al., 2023), daily 5-minute breathwork and mindfulness meditation can both improve mood and reduce anxiety. The researchers found that breathwork may be more effective than mindfulness meditation for improving mood and physiological arousal. In particular, the researchers found that cyclic sighing, a type of breathwork, was the most effective at improving mood and reducing respiratory rate. The researchers also found that both breathwork and mindfulness meditation can reduce negative emotions, such as anxiety.

Also, researchers have shown that deep breathing can help athletes improve their performance by helping them focus, reducing anxiety, and increasing oxygenation (Hunt et al., 2018). Moreover, deep breathing can synchronize brain rhythms, particularly in the alpha range, which are associated with a relaxed yet alert state of mind. Hsu et al. (2020) demonstrated that slow-paced inspiration could organize cortical phase activity in a structured manner, potentially facilitating a state of focused attention and relaxation.

Future studies could help identify a simple deep breathing protocol to be implemented at intervals to modulate the challenge/skill ratio (i.e. when performance is perceived to decrease due to stress), determining specifics of the timeline when this protocol could be used to enter or maintain flow. Post flow, deep breathing could help switch autonomous nervous system states

https://doi.org/10.52202/078697-0003

from relative sympathetic dominance to parasympathetic (rest and digest) state. Cortisol levels and heart rate variability could be measured in a future study comparing a deep breathing post flow group with a control group. Post flow, deep breathing could help to reduce stress and return the heart rate to a relaxed state, especially if the flow activity included a physical dimension, such as sports or other types of embodied performance requiring physical effort.

Conclusion

Flow is a complex mental state, and current research is still being done to uncover all its neurophysiological correlations. Multiple neurocognitive parameters have to coexist to create the optimal conditions for flow: full task engagement, reduced self-referential thinking, a sense of control, and effortless attention. These translate from a neurophysiological standpoint as the activation of the locus coeruleus - norepinephrine system, specific electroencephalogram (EEG) patterns, the interaction of large - scale brain networks, and characteristic cardiovascular and respiratory changes. While there is no shortcut for entering or maintaining flow state, and it is obvious that flow cannot be achieved like flipping a switch, implementing simple tools such as deep breathing might help performers, including musicians, increase their productivity, improve their performance, minimize the effects of physiological arousal and increase their overall well-being. There is sufficient evidence that deep breathing can modulate flow state by enhancing cerebral circulation, promoting autonomic balance, and synchronizing brain rhythms, thereby creating an optimal physiological environment for achieving flow. Future clinical studies could help clarify if and how deep breathing could be implemented to modulate the challenge/skill ratio during flow and as a post flow aftercare technique.

https://doi.org/10.52202/078697-0003

Appendix - Flow Short Scale (Engeser & Rheinberg, 2008)

	not at all			partly			very much
I feel just the right amount of challenge.	○	○	○	○	○	○	○
My thoughts/activities run fluidly and smoothly.	○	○	○	○	○	○	○
I don't notice time passing.	○	○	○	○	○	○	○
I have no difficulty concentrating.	○	○	○	○	○	○	○
My mind is completely clear.	○	○	○	○	○	○	○
I am totally absorbed in what I am doing.	○	○	○	○	○	○	○
The right thoughts/movements occur of their own accord.	○	○	○	○	○	○	○
I know what I have to do each step of the way.	○	○	○	○	○	○	○
I feel that I have everything under control.	○	○	○	○	○	○	○
I am completely lost in thought.	○	○	○	○	○	○	○

	easy								difficult
Compared to all other activities which I partake in, this one is …	○	○	○	○	○	○	○	○	○
	low								high
I think that my competence in this area is ...	○	○	○	○	○	○	○	○	○
	too low				just right				too high
For me personally, the current demands are ...	○	○	○	○	○	○	○	○	○

https://doi.org/10.52202/078697-0003

References

Atkinson, J. W. (1957). Motivational determinants of risk-taking behavior. *Psychological Review, 64*(6, pt.1), 359–372. doi.org/10.1037/h0043445

Balban, M. Y, Neri, E., Kogon, M. M., Weed, L., Nouriani, B., Jo, B., Holl, G., Zeitzer, J. M., Spiegel, D., & Huberman, A. D. (2023). Brief structured respiration practices enhance mood and reduce physiological arousal. *Cell Rep Med. 4*(1):100895. doi: 10.1016/j.xcrm.2022.100895

Brunstein, J. C. (2008). Achievement motivation (H. Heckhausen, Ed.). In J. Heckhausen & H. Heckhausen (Eds.), *Motivation and action* (2nd ed., pp. 137–183). Cambridge University Press.

Csikszentmihalyi, M. (1975). *Beyond boredom and anxiety*. San Francisco: Jossey-Bass Publishers.

Csikszentmihalyi, M. (1990). *The psychology of optimal experience*. New York, NY: Harper & Row.

Csikszentmihalyi, M. & LeFevre, J. (1989). Optimal experience in work and leisure. *Journal of Personality and Social Psychology,56*(5),815–822. doi:org/10.1037// 0022-3514.56.5.815

Csikszentmihalyi, M., & Csikszentmihalyi, I. (1988). *Optimal experience: Psychological studies of flow in consciousness.* Cambridge: University Press.

Colzato, L.S., Wolters, G., & Peifer, C. (2018). Transcutaneous vagus nerve stimulation (tVNS) modulates flow experience. *Exp Brain Res. 236*(1):253-257. doi:10.1007/s00221-017-5123-0

Engeser, S., & Rheinberg, F. (2008). Flow, performance and moderators of challenge-skill balance. *Motivation and Emotion, 32*(3), 158–172. doi:org/10.1007/s11031-008-9102-4

Frishkopf, M. (2003). Authorship in Sufi poetry / موهفم فلؤملا يف رعشلا يفوصلا. *Alif: Journal of Comparative Poetics, 23,* 78–108. doi:org/10.2307/1350077

Hopper, S. I., Murray, S. L., Ferrara, L. R., & Singleton, J. K. (2019). Effectiveness of diaphragmatic breathing for reducing physiological and psychological stress in adults: A quantitative systematic review. *JBI Database of Systematic Reviews and Implementation Reports, 17*(9), 1855–1876. doi:10.11124/JBISRIR-2017-003477

Hsu, S. M., Tseng, C. H., Hsieh, C. H., & Hsieh, C. W. (2020). Slow-paced inspiration regularizes alpha phase dynamics in the human brain. *Journal of Neurophysiology, 123*(1):289-299. doi:10.1152/jn.00624.2019

Hunt, M. G., Rushton, J., Shenberger, E., & Murayama, S. (2018). Positive effects of diaphragmatic breathing on physiological stress reactivity in varsity athletes. *Journal of Clinical Sport Psychology, 12*(1), 27-38. doi:org/10.1123/jcsp.2016-0041

Katahira, K., Yamazaki, Y., Yamaoka, C., Ozaki, H., Nakagawa, S., & Nagata, N. (2018). EEG Correlates of the flow state: A combination of increased frontal theta and moderate frontocentral alpha rhythm in the mental arithmetic task. *Frontiers in Psychology, 9*. doi:10.3389/fpsyg.2018.00300

Keller, J., Bless, H., Blomann, F., & Kleinbohl, D. (2011). Physiological aspects of flow experiences: Skills–demand-compatibility effects on heart rate variability and salivary cortisol. *Journal of Experimental Social Psychology,47*(4), 849–852. doi:org/ 10.1016/j.jesp.2011.02.004

Lao Tzu. (2020). *Tao Te Ching* (John H. McDonald, trans.). London, UK: Arcturus Publishing Limited

.

Li, R., Sughimoto, K., Zhang, X., Wang, S., & Liu, H. (2023). Impacts of respiratory fluctuations on cerebral circulation: A machine-learning-integrated 0-1d multiscale hemodynamic model. *Physiological Measurement, 44*(3). doi:10.1088/1361-6579/acc3d7

Maslow, A. (1954). *Motivation and personality.* New York, NY: Harper & Brothers.

Moneta, G. B., & Csikszentmihalyi, M. (1996). The effect of perceived challenges and skills on the quality of subjective experience. *Journal of Personality, 64,* 274–310. doi: 10.1111/j.1467-6494.1996.tb00512.x

Murch, W. S., Ferrari, M. A., McDonald, B. M., & Clark, L. (2020). Investigating flow state and cardiac pre-ejection period during electronic gaming machine use. *Frontiers in Psychology*. doi:10.3389/fpsyg.2020.00300

Noble, D. J., & Hochman, S. (2019). Hypothesis: Pulmonary afferent activity patterns during slow, deep breathing contribute to the neural induction of physiological relaxation. *Frontiers in Physiology.* doi:10.3389/fphys.2019.01176

Peifer, C., Schulz, A., Schächinger, H., Baumann, N., & Antoni, C. (2014). The relation of flow-experience and physiological arousal under stress — Can u shape it? *Journal of Experimental Social Psychology, 53,* 62-69. doi:10.1016/j.jesp.2014.01.009

Peifer, C., Schächinger, H., Engeser, S., & Antoni, C .H. (2015). Cortisol effects on flow-experience. *Psychopharmacology, 232*(6):1165-73. doi:10.1007/s00213-014-3753-5

Russo, M. A., Santarelli, D. M., & O'Rourke, D. (2017). The physiological effects of slow breathing in the healthy human. *Breathe, 13*(4), 298–309. doi:10.1183/20734735.009817

https://doi.org/10.52202/078697-0003

Ullén, F., de Manzano, Ö., Almeida, R., Magnusson, P., Pedersen, N., Nakamura, J., Csikszentmihalyi, & Madison, G. (2012). Proneness for psychological flow in everyday life: Associations with personality and intelligence. *Personality and Individual Differences, 52*(2), 167-172. doi:10.1016/j.paid.2011.10.003

Van der Linden, D., Tops, M., & Bakker, A. B. (2021). Go with the flow: A neuroscientific view on being fully engaged. *The European Journal of Neuroscience. 53*(4), 947-963. doi:10.1111/ejn.15014.

Human as a Multidimensional Harp
A Convergence between Therapeutic Music and Vibroacoustic Harp Therapy

Marianne Gubri

Abstract

Each human being is a complex structure of different layers of music and sound. With their proven impact on response mechanisms of the brain and the body, music and sound may contribute to harmonising different aspects of human existence such as physical, emotional, cognitive, and spiritual spheres.

Harp therapy provides a sonic cocoon; its positive effects have been widely described in scientific literature. Vibroacoustic Harp Therapy (VAHT) provides a deeper frequency massage throughout the entire body, which may reduce pain, anxiety, and stress.

Based on a private VAHT study, through this presentation the researcher will propose a convergence of these approaches. It will be possible to draw an indicative map of resonant frequencies in the body and to investigate how to create balance and harmony. Some tones may also touch specific blocks, which resonance may induce a release of emotions, potentially balanced with modal and rhythmic changes.

Keywords: Harp Therapy, Vibroacoustic, Frequencies; Entrainment, Modes

https://doi.org/10.52202/078697-0004

Introduction

Therapeutic music and vibroacoustic therapies have been focused on the frequency resonances in the body. A specific technique, vibroacoustic-harp therapy (VAHT), has led practitioners to understand its positive effects on pain, muscular rigidity, and cardiac variability; clients respond to one or more sessions of VAHT with increased relaxation, new awareness of specific areas of the body, stimulation of the immune system, better sleep, and a sensation of being grounded and energized. This occurs especially with patients affected by multiple sclerosis, cancer, anxiety, pain, fibromyalgia, and/or dementia.

Based on an independent study on 40 clients between July 2022 and March 2023, this researcher will highlight new directions for further studies on frequencies and resonance, in particularly: how different parts of the body react with different frequencies and how they are organized. It will then be possible to draw a dynamic map of frequencies and to link these frequencies to other vibrational therapy models. Moreover, we will understand how these techniques are not only related to physical awareness, but also to emotional release and mental imagination; this research may lead to new considerations on therapeutic music and sound healing practices. The aim of these techniques might be to manage pain and stress, and to allow lateralisation between the right and left brain, homeostasis, and self-awareness.

Vibrational Therapies

Vibroacoustic therapies (VAT) focus mainly on the resonance of live or recorded frequencies with the tissues of the body, leading to physiological processes via a specific device (bed, chair, pillows, wristbands...). These methods have been studied in-depth by Skille (2017), who, among others, worked mainly on low frequencies (30-120 Hz) with autistic children. VAT is now globally used as a valid technique to manage pain, stress, insomnia, and anxiety, as well as in improving circulation, brain-heart coherence and sleep in spas, schools, hospitals, and clinical settings.

Vibroacoustic Harp Therapy (VAHT®), developed by Williams (2005, 2013), an American harpist and psychologist, not only uses low frequencies but additionally all the range of the harp, which, for a concert harp, extends from Cb0 (30.9 Hz) to Gb7 (2960 Hz). The instrument is connected via a microphone through an amplifier to a device (a massage bed or a chair) implemented with specific transducers. The principle of vibroacoustic is that we not only feel sound with our ears but also with our entire body, which becomes a sounding board (Dallas-

https://doi.org/10.52202/078697-0004

Fenney, 2004). Vibrotactile perception has been described by Melzack as the Gate Control Theory (Williams, 2005) which explains how non-painful sensations can reduce painful sensations. Pacinian corpuscles, the mechanoreceptors located in subcutaneous and connective tissue, are declared responsible in detecting mechanical stimuli of a vibratory nature, whose transmission would be able to interfere with the transmission of painful signals (Mezlack, 1975; Patrick,(1999). Moreover, a proposed mechanism of action of vibroacoustic therapies is brainwave entrainment that would synchronize with rhythmic inputs on a sensory level. A typical VAHT session lasts from 30 to 60 minutes, with a first part dedicated to the body frequencies scan, and a second part filled with musical improvisations and relaxing music (Gubri, 2022).

On a sample of 40 clients, all of them were able to feel the vibrations of the concert harp frequencies (47 strings) in their bodies through a massage bed device. All of them had the perception of a decrease in tension/pain by at least two points on a scale from 1 to 9. Ten people had emotional release and twenty experienced an oneiric process (visualization of landscapes, fractals, images...)

Resonance

VAT and VAHT are described as vibrational therapies since their focus is mainly sound and sound vibration in the body. It also engages another interesting phenomenon which is called resonance. According to Beaulieu (2010), "Resonance comes from the Latin verb *resonare,* meaning to "return to sound. It means to sound and resound as in an echo." (p.152). Sympathetic resonance is an event in which a frequency emitted by a sonic source meets an object or a body's resonance frequency creating an increased amplitude compared to another body with a non-resonant frequency (McMakin, 2017).

In a VAHT session, sympathetic resonances is easily understood since each frequency resonates in a specific part of the body, which may lead to understanding that each part of each body has a specific resonance frequency. The first part of the session is dedicated to "scanning" the body with all frequencies, string per string. The majority of the frequencies are physically felt within a range between 30 to 500 Hz. Few people feel the resonance in the body over 500 Hz, although the acoustic sound of the instrument through normal auditory mechanism can still be heard.

Each frequency "hits" a specific place in the body, an organ, a bone, an area, sometimes more than one area (for instance an A1 will resonate in the feet and in the lower back of a client).

https://doi.org/10.52202/078697-0004

Some frequencies do not seem to have a physical response from the client. All frequencies are delivered one by one, while the client is invited to verbalize his sensations, giving feedback to the practitioner who records the results on a form. This kind of feedback relies on the fact that the client has a minimum of body self-awareness (proprioception) and the capacity to express themselves verbally. We can also have more refined feedback, looking at the breath rhythm or at micro-modifications of their posture and /or their face.

The first interesting observations are that all clients feel surprised by a new awareness of their body: sound is vibrating from the inside, in a real sonic massage. Vibration is described by the clients as inner movement, warmth, chills, waves that move inside the body, light, caress, bubbles, energy...Most clients express their curiosity on how sound can touch the tissues so profoundly, and how they might become conscious of a pellicular organ (for instance the weight, space, warmth of the heart). Five out of 40 clients felt the sound outside their body; this new sensation may open research on the perception of the electromagnetic field by the client him/herself through sound. Some frequencies might be felt at a very specific and pleasurable point (on the coccyx for instance) while for others it might be felt more widespread. Some other frequencies are expressed as not so pleasurable, and they may lead to emotional release, fear, or rejection. Generally, after the session, the clients describe their body as heavier, more relaxed, and more grounded, while self-consciousness is lighter, freer movement, and their mood is generally improved (Miller, 2011).

Another observation is that tension, trauma, and emotional blocks might be released through sound without any additional pain. The activation of parasympathetic system may induce a deeper relaxation, eventually leading to sleep, and the neuro-flow is activated, while a new flow of information is carried to the thalamus cortex (Valone, 2003). During a VAHT session, Client P, a young man who had undergone several leg operations due to a preterm birth, was positively surprised to feel the same relaxation of tissue adhesions in his right leg that he had had during kinesiotherapy sessions, but without pain.

Absence of vibration response of a frequency in a part of the body is often connected to a specific pain or illness in the same place. With more than one session (on a weekly basis), we can observe how clients become more and more sensitive. They are more aware of several places in the body. They may find that the same frequencies give them more resonances in the body. They may also feel that the frequencies present in part of the body where they were in pain or in tension are now slightly relaxing. Client D. was not able to feel the resonance of the frequencies in his right arm that was hurting, however, he was able to feel the frequencies in the left arm. After some sessions, he could feel in his right arm the same frequencies he felt in his left arm. VAHT may open doors to resonance, body awareness, and balance.

It is not rare, without any specific pain, to also see unbalances between the left and right part of the body. We will discuss possible causes and consequences of such perception on a vertically divided body awareness later. The most interesting observation is that there is not one common frequency for everybody. Every frequency might resonate in a different place for every client, which means that there is not a single identical resonant frequency for every organ, bone, body etc, but every single frequency is useful for a single person. Resonance depends on weight, liquid/material composition, and heat tension, which may vary considerably from one person to another.

On the contrary, we can measure how, during different sessions, the same resonance pattern will show up for a specific client which may lead to understanding their sonic individual map. Goldman (2008) created the formula "We are all unique vibratory beings" in his book *The 7 Secrets of Sound Healing* (p. 47). This sonic signature reveals some consistent data: pleasurable frequencies, absence of vibrational response, same resonances with identical frequencies. We all have typical resonant frequency tendencies.

Frequency Mapping

Frequencies and frequency mapping is one of the most intriguing questions in this historical period of great interest in sound healing and therapeutic music. The wide curiosity in oriental spirituality and self-awareness practices may open new perspectives to these questions, even if we might avoid a quick and superficial association between frequencies and parts of the body. Today, several informative volumes are being published (Beaulieu, 2010; Maman, 1997; McKusick, 2014, Pierce, 2009). These researchers help answer questions such as: What are the connections between the organs? Is our body subject to a sonic somatotopy? Is there a connection between sonic physiologic resonances and psychological aspects?

While mapping the different frequencies, some consistent data emerged. Low frequencies (32-130 Hz) resonate in the lower part of the body (feet to abdomen), while higher frequencies (130-523 Hz) resonate in the higher part of the body (thorax to head). As referred by Gibson (2022, p. 56), we can observe a "hierarchy of frequencies," even if regularity and

https://doi.org/10.52202/078697-0004

homogeneity are not on a constant progression. The different frequency perceptions are widespread and fluid like the differences from red to orange on a natural optical rainbow. Furthermore, the data take-over highlights an organisation of the frequencies. A further observation indicates a division in 4 different octaves from approximatively 32 Hz to 523 Hz. The perception of vibrations of the fifth octave (above C5 – 523 Hz) is rather rare in the clients of this study. These division in four octaves may be read as follows, (frequency charts are based on an instrument tuned at A = 440 Hz):

- From C1-C2 (32,703-65,406 Hz) resonances are mainly felt in legs and feet.
- From C2-C3 (65,406-130, 81 Hz), resonances are mainly felt in hips, belly, and abdomen.
- From C3-C4 (130,81– 261,63 Hz), resonances are mainly felt in heart, lungs, back, arms.
- From C4- C5 (261,63- 523,25 Hz), resonances are mainly felt in throat, head, ears, and eyes.

While highlighting the more pleasurable notes on the frequency map, it has emerged that these notes are often in an octave succession. For instance, a sequence of A's on three different octaves has been discovered on client L., a patient who really enjoyed this note, physically and acoustically. In our modern conception of music, we are used to starting an octave from the C note. As a matter of fact, it could be from any other note on the musical scale, as in traditional Indian music where notes are not linked to a specific frequency but to a position of the scale. This observation is linked to the detection of the individual tone of resonance, the tone we mainly use in our natural spoken voice (Dietzel, 2004). With this division, we have different octaves which start from the tone that pleases us more and resonates more in our bodies and voices. Each client might then have their own frequency progression in octaves based on their natural resonant tone.

This organised hierarchy has been described as a sonic somatotopy by Aucher (1977), a French singer and inventor of the *psychophonie*, a method of vocalization along the spinal cord to detect blocks and holes with the natural resonances of the voice. In the researcher's own body scan, called *cliché des sons*, Aucher has highlighted four octaves of resonance in the vertebrae on the spinal cord, a model which Aucher linked to acupuncture before registering it at the *Académie des Sciences* in Paris, now widely spread all over Europe through her students and academies. This division on the vertebral column is very similar to the Governor Vessel, a vessel of subtle energy studied in acupuncture, as Aucher discovered thanks to discussions with traditional Chinese doctors in the last decades of Aucher's life.

Psychosomatic Resonances

The emergence of emotional release and specific imaginative material during VAHT session may allude to a specific connection between somatotopic and psychological activities. While psychosomatic medicine, thanks to the works of Shelton (1952), is now finally well accepted in the field, it would be interesting to understand how vibrational response in the physical body might be linked to emotional content, as a method of diagnosis or interpretation of discontinuity in vibrational somatotopic response. For instance, in several cases, I observed clients having several differences in perception between left and right part of the body, which might be linked to the right and left brain. An absence of resonance of a certain point of the body may explain physical tension and pain, but also a psychic disharmony to a part linked to this area.

De Souzenelle wrote an essay (1998) on the meaning of illnesses and how the symbolical understanding of a specific pain might give further awareness to the client with the objective to recover and gain harmonic health and wellness. Illness becomes a cure to encourage psychological and spiritual awareness. The more an organ is stressed, the more it is at risk of carrying somatization:

> The organism is not able to disguise between what is fear and reality, between the physical and the psychic; it somatises in the same way. Faced with a difficulty, the organism reacts by mobilizing its means of defence: it is the stress, in its most positive aspect. Physical illness might be a response to psychic traumas; mute frequencies on the body could then be linked to a symbolic response to a traumatic event or an emotional disease, event before the existence of a specific, possibly chronic physical illness. (De Souzenelle, 1998, p.72).

VAHT and resonance therapies might then be understood as a tool for illness prevention and balance and wellness maintenance.

Aucher (1977, 1983) has linked sympathetic resonances outside of the body to a psychological activity. As a researcher, Aucher finally described one's full-resonating sonic-human being with a seven octaves model, from the lowest octaves corresponding to the lowest parts of the body to the highest octaves corresponding to the highest parts of the body:

- four octaves for the "physical body" subdivided in one octave for motor and rhythmic activity, one octave for vegetative and sexual activity, one octave for affective and pulmonary activity, one octave for nerve and sensorial activity. These divisions can similarly be observed and matched through VAHT frequency mapping.

https://doi.org/10.52202/078697-0004

- One octave for other "ethereal material" - which, in VAHT studies, might be linked to the sensation of feeling the sound out of the body, often described as the biofield (McKusick, 2014) or electromagnetic field (Milroy, 2013).
- One octave for the "astral body", connected to the expression of feelings.
- One octave for the "glorious body," which might be linked to oneiric materials and purpose of life.

Such a division in layers of existence is not new and can easily be linked to the yogic tradition (Dale, 2009) which calls them the *koshas* or *sheaths* ordered from the denser matter (the physical body) to the more ethereal or spiritual state (the etheric, astral, mental, and causal bodies). These different bodies seem organised according to their vibratory level and state of consciousness. To cite Gerber (2001),

> The difference between the physical matter and etheric matter is only a difference of frequency. It is an acknowledged principle within physics that energy of different frequencies can coexist within the same space without destructive interaction [...] The energetic matrix of the etheric body, or holographic energy-field template, is superimposed upon the structure of the physical framework. [...] Bodies of higher energetic frequencies are interconnected and in a dynamic equilibrium with the physical body. (p.120)

This human multidimensional anatomy might be seen as a harp and its several strings can combine resonance and synchronicity through chords and consonant intervals to maintain or regain harmony.

Metamerism

Auriol, a French psychiatrist and father of *Psychosonique* science (1991), has discovered that some bodies are segmented in several horizontal zones, called metameres, as clearly represented in the worm family. Regarding the human body, the yogic tradition has described the human energetic system as a superposition of seven chakras, or energetic centres (they may differ in number in other cultures, as we can find five chakras in Tibetan culture, sometimes 12 or even more in other traditions). According to ancient schools like Nada Yoga (the Yoga of Sound) and Mantra Yoga, these chakras are linked to a specific resonant point in the physical body (the endocrine system with its hormones), they represent a specific psychic awareness (seat of consciousness), and they may be linked to a colour frequency (from the rainbow optical spectrum) and an audible frequency (from the natural C major scale). The most represented and well-known model (Goldman, 2008) is the one which connects the seven chakras to the seven notes of the scale (C D E F G A B – we just need to remember that the notes in the Indian tradition are mobile and not fixed frequencies).

We can now give another interpretation to this model, in which each tone of each chakra may represent an octave in the physical body: C for the first octave, D for the second octave, and so on. The Armenian philosopher Gurdjieff (Ouspensky, 2001) described this phenomenon as the lateral octaves, the seat of inner evolution and consciousness. Moreover, he emphasized the comprehension of the change of an octave as a double frequency and double speed, which correspond to a higher state of being. From the motor plane to the vegetative plane, there is a difference of doubled vibrations; from the emotional to the mental plane, another difference of doubled frequencies. We can now understand our human body as a multidimensional harp, which connects body awareness, emotions, thoughts subtle energy through invisible strings that sound healing and vibroacoustic can reveal.

In several traditions, the birth of the harp was linked to an extraordinary mythical vision. Mythology often records of ancient lyres made of seven strings (Williams, 2005): Hermes, the messenger of the gods in ancient Greek tradition, was often known as the inventor of the lyre, made of seven strings from a tortoiseshell and gut found in a cave. In other traditions, the seven strings of the lyre were linked to the Seven Sisters, the seven stars of the Pleiades, thought as divine singers and dancers, according to a tradition that linked audible sounds to the doctrine of the cosmic harmony called the music of the spheres. Hermes gave his lyre to his brother Apollo, God of light, music, medicine, and protector of the Muses; he was also the holder of the caduceus, which is still a symbol of many healthcare organizations today and which several authors (Chaudhary, 2020, Goldman, 2008) have linked to the yogic tradition of the three *nadis* (energetic channel or sympathetic, parasympathetic, central nervous systems). Several ancient poets describe the lyre as a symbolic representation of the cosmos (Molina Moreno, 2008). In India, a yogic tradition links the *veena*, a sort of lute (formerly an ancient harp) to the physical body: each string corresponds to a chakra and the frets on the neck of the instrument match the vertebrae of the back while the resonator is associated to the head. In such a tradition, playing the *veena* is synonymous of harmonizing the physical and energetic body, according to the therapeutic knowledge of *ragas* (modes), *svaras* (tones), and *tala* (rhythm). Lying on a massage bed, a VAHT client is harmonising their seven endocrine systems, their different points of energy and body awareness; as therapeutic musicians and sound healers we can relate to an ancient

 https://doi.org/10.52202/078697-0004

tradition of healing music to continue, through modern science and concrete practice, a millennial knowledge.

Musical and Sonic Considerations

As therapeutic musicians, we can investigate on how, after a sonic body scan, we can improve our sessions with specific sequences, harmonies, intervals, rhythms, frequencies, touch, which are among the different variability of sound parameters. I will share some of the experiences I have met during the time-lapse of a nine-month VAHT practice and some complementary considerations given by other authors.

- All frequencies influence our body.
- Low frequencies (30-65 Hz) can be exceptionally grounding and relaxing to the physical body. A large documentation of this effect is available through VAT studies.
- Medium frequencies (65 – 260 Hz) might be useful in connecting with the emotional part of our clients.
- Higher frequencies (260-523 Hz) might be useful to connect with their mental (brain) and energetic dimension (electromagnetic biofield).
- Frequencies that do not resonate on the physical body may be played anyway to open this part of the body which might be tense, in pain or not often used. It is a good way to balance the resonant body.
- Frequencies that are hitting a painful point could be played with a very delicate touch and with low intensity. Avoiding a direct and intense touch might help to relax the client and, at the same time, it might be interesting to approach those frequencies from other nearby frequencies which will surround the "painful frequency," and then gradually approach it, creating a forced resonance.
- Some frequencies really hit or please the clients at a very precise point: I often play these frequencies at the beginning of the relaxing session and create a harmonisation based on chords sequences, like the natural harmonic series.
- Harmonic sequences can lead to inner harmony: beautiful series of arpeggios, intervals, and chords can be done following the natural harmonic series; it has without doubt a connection with the harmonisation of the physical, emotional, mental, and spiritual parts (Gardner, 1990, pp. 102-117). In the lower octave, I would play larger and more open intervals, like octaves and fifths. For the central octave, thirds and sevenths intervals suit better while melodies and scales are perfect on the higher octaves, in a progression that reflects the constant and progressive restriction of intervals in the harmonic series.
- Modes have specific emotions and energetic dimensions: from ancient Greek scales to Gregorian modes, from Indian *ragas* to Middle Eastern *Maqam*, all musical traditions have given emotional and therapeutic attributes to every type of musical scale.

- Each mode can be determined with its emotional level in an ascending spiral from the minor and more introspective Locrian, then Phrygian, Aeolian, and Dorian, to the more outreaching and major modes like the Mixolydian, Ionian and finally Lydian mode.
- While giving a vibroacoustic harp therapy session, I often switch from a deep introspective mode to a more joyful and light mode at the end of the session, for instance Aeolian or Dorian to Mixolydian or Ionian.
- As for the rhythmic delivery of the single notes, I will ask the client if they prefer to relax (mostly) or to dynamize themselves during the session. Typically, I will follow a flow conduction of the session, with a more relaxing part at the beginning and more dynamizing at the end. During the beginning of the improvisation, I will follow the biorhythms of the client, especially the rhythm of their breath and I will try to retrace the same tempo with the notes. Thanks to entrainment and natural synchronization of the body, the heart, and the brain waves with the breathing rhythm, it is possible to slow down the biorhythms of the client and to convey a more relaxing pace. Deep relaxation through sound has a proven positive effect on the activation of the vagus nerve, in control of the rest response, digestive function, stress, and pain management.

Conclusions

The observations realized during this study can lead to different hypothesis. The visualization of the frequencies response on the physical body helps us to understand blocks, tensions, and traumas and eventually to promote a further investigation and a diagnosis from a medical professional. Physical and psychological awareness are deeply increased during VAHT sessions and can indicate a better stress and pain management from the patient's point of view. Differences in resonances from the left and right part of the body can give interesting suggestions to balance body functions and brain lateralization, which, as a result, may open doors to personal creativity, intuition, and productivity. A deeper observation of different points where sound resonances are present or absent may open a discussion on psychological identification, thanks to a deeper focus on morphopsychology and symbolism of the different parts of the body on the subtle plane (emotional, energetic, and spiritual planes).

VAHT has been confirmed as a technique to improve stress management, tension, lateralization on the left and right brain and left and right part of the body, coherence between brain and heart, regulation of heart variability, motor and space training, body and emotional awareness, and homeostasis in order to regulate harmony between different biological systems. We can also imagine a near future where vibrational and vibroacoustic therapies will be recognized as a useful tool for prevention and maintenance of health and wellness. A quote from

https://doi.org/10.52202/078697-0004

Maman (1997), *The Body as a Harp*: "...observing and harmonizing the different frequencies of the body is the beginning of understanding the multidimensional human being layered with different qualities of vibrations, awareness, and manifestations" (p.33).

References

Aucher, M. -L. (1977). *L'homme sonore*, Epi.

Aucher, M. -L., (1983). *Vivre sur sept octaves*, Résonances.

Auriol, B., (1991). *La clé des sons, Eléments de psychosonique*, Erès.

Beaulieu, J., (2010). *Human Tuning, Sound Healing with Tuning Forks*, BioSonic.

Butler, C., Butler, P. (1997). Physioacoutic therapy with cardiac surgery patients. In Wigram T., and Dileo, C. eds., *Music Vibration and Health*. Jeffrey, 197-204.

Chaudhary, K. (2020) *Sound Medicine,* Harper Wave.

Dale, C., (2009). *The Subtle Body, An Encyclopedia of Your Energetic Anatomy*, Sounds True.

Dallas-Fenney, S. (2004). Pilot study demonstrates positive effects of vibroacoustic harp therapy on heart rate variability. *The Harp Therapy Journal, 9*(1), 1-9.

De Souzenelle, A. (1998). *Le symbolisme du corps humain*, Dangles.

Dietzel, G. (2004). *The individual Sound of the Voice, Its Message, Its Action*, Druckerei.

Gardner, K. (1990). *Sounding the Inner Landscape*, Element.

Gerber, R. (2001). *Vibrational Medicine*, Bear.

Gibson, D. (2022). *The Complete Guide to Sound Healing*, s.n.

Goldman, J. (2008). *The 7 Secrets of Sound Healing,* Hay House.

Gubri, Marianne. (2022). Expanding VAHT in Europe. *Harp Therapy Journal, 27*(2), 5-15.

Maman, F. (1997). *The Body as a Harp: Sound and Acupuncture*, Tamado.

McKusick, E. (2014). *Tuning the Human Biofield,* Healing Arts Press.

McMakin, C. (2017). *The Resonance Effect, How Frequency Specific Microcurrent Is Changing Medicine*, North Atlantic Books.

Mezlack, R. (1975). Prolonged relief of pain by brief, intense transcutaneous somatic stimulation. *Pain, 1*(4), 357-373.

Miller, E. M. (2011). *Bio-guided Music Therapy, A Practitioner's Guide to the Clinical Integration of Music and Biofeedback*, Kingsley.

https://doi.org/10.52202/078697-0004

Milroy, M., (2013). Human biofield measurements before and after VAHT. *The Harp Therapy Journal*, *18*(4), 10-14.

Molina Moreno, F. (2008). The Pleiades or the First Cosmic Lyre. *Hyperboreus*, *14*(1), 28-38.

Ouspensky, P. D. (2001). *In search of the Miraculous,* Tapa Blanda.

Patrick, G. (1999). The effects of vibroacoustic music on symptom reduction: inducing the relaxation response through good vibration. *IEEE Engineer Med Biol.*, 97-100.

Pierce, P. (2009). *Frequency, The Power of Personal Vibration*, Simon and Schuster.

Shelton, W.P. (1952). Psychosomatic medicine, *Journal of Medicine*, *48*(3), 121-124.

Skille, O., Palmer, R., Lahiten, R., Ojala, S. (2017). Feeling Vibrations from a Hearing and Dual-Sensory Impaired Perspective, *Music & Medicine, 9*(3), 178 – 183.

Valone, T. (2003). *Bioelectromagnetic Healing, A Rationale for its Use,* Integrity Research Institute.

Williams, S. (2013). VAHT offered in a primary care setting C An interview with Warren Ross, M.D. *Harp Therapy Journal*, *18*(2), 1-15.

Williams, S. (2013), Vibroacoustic Harp Therapy in Pain Management: Soothes body, mind, and soul, *Practical Pain Management*, *13*(2), 56-59.

Williams, S. (2005). *Good Vibrations, Principles of Vibroacoustic Harp Therapy*, Musiatry.

Survey of University Acting Majors Regarding the Effects of Singing Training on Confidence and Skills

Brian Manternach & Jeremy N. Manternach

Abstract

A three-semester course for university acting majors had a three-fold focus: (1) performing songs, (2) music theory, and (3) vocal pedagogy (physiology, technique, wellness). After the course had been taught for four years, 59 students assessed which curricular elements were most useful while taking the course(s) and once they had completed the course(s). As previously published, while taking the course(s), the highest number of participants indicated that learning and performing songs was the most useful, and vocal pedagogy least useful. However, perceptions inverted after taking the courses, with vocal pedagogy rated most useful and learning and performing songs least useful. Of those who participated in auditions or stage productions that required singing, more than 90% reported the courses were extremely or somewhat helpful in building their confidence and more than 95% reported the courses were extremely or somewhat helpful in building their skills.

In this investigation, we examined previously unreported data from the open-response questions: (1) Overall, what do you feel is valuable about the singing training you took as you were training to be an actor? (2) Do you have any other comments or reflections on the course? Forty-seven (79.7%) participants mentioned improvements in confidence and/or skills in their answers. Of those, 57.4% mentioned improvements in confidence or used a synonymous phrase compared to 46.7% who specifically mentioned improvements in skills. Additionally, 31.5% indicated improvements in confidence or used a synonym without mentioning improvements in skills, while only 20.4% mentioned improvements in skills without mentioning confidence.

The increased feelings of competency may have increased motivation. Similarly, the coursework seemed to address multiple components of Bandura's self-efficacy theory. Increased domain-specific self-efficacy (e.g., singing performance) has been linked to general self-efficacy; and high general self-efficacy has been linked to psychological well-being.

Keywords: Performance Skills, Confidence, Self-efficacy, Psychological Well-being, Theatre, Singing

https://doi.org/10.52202/078697-0005

Introduction

Researchers have examined the connection between participation in music and an increase in self-confidence among those in specific populations. In one study, adolescents who have been sexually abused found an increase in self-esteem and self-confidence after music therapy interventions (Clendenon-Wallen, 1991). In a second study, multiple sclerosis patients experienced an increase in self-esteem after multiple sessions of music therapy (Aldridge et al., 2005). In a third, Taiwanese undergraduate students experienced increased confidence after 20 hours of music therapy (Wu, 2002). Finally, Pavlicevic et al. (2014) posited that "therapeutic musicking" provides young adults with ongoing opportunities for experiencing confidence and self-esteem.

In other studies, researchers found that singing, in particular, may also lead to increased self-confidence, among other benefits (Hillman, 2002; Lestari & Handayani, 2017). In one study, adult subjects experiencing bereavement found improvements in their self-esteem, as well as in their sense of self-efficacy, after joining a choir (Fancourt et al., 2022). Similarly, other sources have tied theatre education and theatre practices to increased confidence and self-esteem (Savatgy, 2022; McCammon, 2012). In fact, certain theatrical programs and curricula have even been structured with a central goal of building participants' self-confidence (Bogwardt, 2010; Horne, 2017; Côté, 2011).

In this study, we combined elements of music, singing, and theatre to examine the impacts on self-confidence. Specifically, the purpose of this study was to determine the effect of singing training on confidence and skills in students in an actor training program housed within a university department of theatre.

Method

Background

In a previous study (Manternach & Manternach, 2019), we investigated a three-semester singing course curriculum that was structured according to three specific student learning outcomes: (a) build performance skills by learning and performing songs, (b) build basic music theory/sight reading skills, and (c) learn basic vocal pedagogy (e.g., anatomy, physiology, vocal technique, vocal wellness).

After the course had been taught for four years, we designed an online survey to measure the perceived effectiveness of the curriculum, collecting input from both currently enrolled students who had completed at least one semester of the three-course sequence and graduates of the curriculum. Among the questions, participants were asked to rate which of the three elements of the courses (performing songs, music theory, or vocal pedagogy) they believed were most useful while (a) they were taking the course(s) and (b) once they had completed the course(s). In addition, they reported how well the course prepared them for professional pursuits. Specifically, they indicated how well the course helped build their confidence and their skills for subsequent auditions and productions that required singing.

Fifty-nine of the 72 eligible students who had taken at least one course in the three-course sequence completed the survey (81.9% response rate). These students had taken one of the courses less than one year (n = 21, 35.6%), between one and two years (n = 24, 40.7%), between two and three years (n = 11, 18.6%), or more than three years (n = 3, 5.1%) prior to completing the survey.

The results indicated that while taking the course(s), the highest number of participants (n = 26, 44.1%) reported believing that learning and performing songs was the most useful element of the course, followed by music theory (n = 23, 39.0%) and vocal pedagogy (n = 10, 17.0%). However, perceptions inverted when subjects reported their impressions after taking the courses, with vocal pedagogy rated most useful (n = 24, 40.7%), followed by music theory (n = 32.2%) and learning and performing songs (n = 16, 27.1%). This trend increased as students were further removed from the course.

Of those who participated in auditions (n = 43, 72.9%) or stage productions (n = 29, 49.2%) that required singing, more than 90% reported the courses were extremely or somewhat helpful in building their confidence and more than 95% reported the courses were extremely or somewhat helpful in building their skills.

Open-Response Questions

In the current investigation, we examined previously unreported data from the two open-response questions at the end of the survey that further explore relationships between the singing course and the development of confidence and skills. The open-response questions were (1) Overall, what do you feel is valuable about the singing training you took as you were training to be an actor? (2) Do you have any other comments or reflections on the course that you would like to share? First, we looked at the number of times participants used the words "skills" or "confidence" in their responses. We then coded particular words that we deemed to be

https://doi.org/10.52202/078697-0005

synonyms of "skills" or "confidence." Finally, we noted the number of times the comments indicated that the comment was positively or negatively worded.

Results

Fifty-four of the 59 participants (91.5%) completed the open-response questions. Forty-seven of those 54 participants (87.0%; 79.7% of all survey participants) mentioned improvements in confidence and/or skills in their open-response answers. Twenty-four (44.4%; 40.7% of survey participants) specifically mentioned a boost in confidence, using phrases such as "It helped me gain confidence performing in an area I don't normally perform in" and "I now feel more confident in my singing." By comparison, nine participants (16.7%; 15.3% of survey participants) specifically mentioned a boost in skills, using phrases such as "I picked up skills and vocal techniques that I still use to this day" and "The class also helped out with keeping my singing skills fresh."

We then examined words or phrases that are synonymous with "confidence." With this approach, an additional seven participants fell into the confidence category, making such comments as "[It] helped [make] singing in front of people less scary" and "[T]he most valuable thing I learned is to trust my own voice." When including synonymous words and phrases for "skills," an additional 16 participants were added to the "skills" category. These comments included direct synonyms (e.g., "abilities") or specific skills (e.g., "Helped me sing in a way that's healthier for my voice" and "It helped me...understand where I'm breathing from.").

Therefore, 57.4% of those who responded to the open-ended questions (52.5% of all survey participants) specifically mentioned improvements in confidence or used a phrase that is synonymous with confidence. By comparison, 46.7% of respondents (42.4% of survey participants) specifically mentioned improvements in skills or used a phrase synonymous with skills.

When considering those who indicated either confidence or skills, but not both, 17 respondents (31.5%; 28.8% of survey participants) mentioned confidence or a synonym for confidence without mentioning skills or a synonym for skills. By comparison, 11 respondents (20.4%; 18.6% of survey participants) mentioned skills or a synonym for skills without mentioning confidence or a synonym for confidence.

It is worth noting that five of the 59 survey participants (8.5%) gave an open-response answer that was unrelated to skills and/or confidence. There were also two responses that we could not determine to be related to skills or confidence or both and were, thus, not included in the percentages above: "It taught me that I can be in musicals," and "The training I received really helped me find my voice." Three of the responses were related to confidence and/or skill, but were negative: "It applied in situations where I had to sing for a show but I don't personally feel confident enough to audition for a musical," "...as a student in the actor training program who cannot sing, it was frustrating to be forced to take three semesters," and I was always feeling bad about my voice, and by the end the structure of the class made me lose confidence in myself as a singer."

It is also noteworthy that, although the survey was measuring self-perception, three participants noted improvements in classmates. One participant noted "I was already confident and comfortable singing in front of people...But for others in my class that had less experience, I think it was important for them to gain confidence and become comfortable singing in front of people." Another participant noted that they "...really enjoyed seeing my classmates' skills and confidence grow." A third indicated that "...a lot of what I learned was supplemental. However, I really enjoyed learning how other people learn to sing in a class." Although these comments indicate a boost in confidence and skills, since they are not related to individuals' self-perceptions, they are not included in the percentages above.

Discussion

As stated previously (Manternach & Manternach, 2019), the element of the curriculum that was perceived to be the most useful changed over time. Initially, respondents felt best served by learning and performing songs. After taking at least one semester of the curriculum, however, more respondents saw greater value in vocal pedagogy and music theory. One possible explanation for such a shift is a realization that basic music literacy and vocal technique skills would allow them to learn songs and negotiate technique independently outside of class, making the in-class activities of learning and performing songs less necessary.

Another possible explanation could relate to a "fixed" (Dweck, 2017) or "talent" (Demorest, 2018) mindset versus a "growth mindset." Anecdotally, many students in the class identified as actors, not singers, and seemed to initially have adopted a fixed/talent mindset about singing, believing that individuals are either born with or born without singing talent. Perhaps an understanding of vocal pedagogy and the function and technique behind singing revealed it to be a learnable skill. Some students appeared to change their mindset, as evidenced

https://doi.org/10.52202/078697-0005

by open-response answers such as "I think that a lot of people are scared of singing but [this] class has helped curb that fear for a lot of people" and "[I]t gave me so much confidence. Anyone can sing is a lesson well learned."

Relating to the confidence versus skills portion of the open-response questions, since participants were asked directly about perceived improvements in confidence and skills earlier in the survey (essentially priming them to consider these factors), it is unsurprising that they are mentioned prominently in the open-response answers. However, the fact that improved confidence, or phrases synonymous with confidence, received greater—though not overwhelming—mention than improved skills, or phrases synonymous with skills (or any other area of improvement, for that matter), may be an indication that the greatest overall benefit of the class was related to an increase in confidence.

An increase in confidence may also be linked to motivation and basic psychological needs. Ryan and Deci (2000) have discussed the three basic components of motivation: competence, relatedness, and autonomy. Given the number of participants who reported on increased confidence and skills, it is clear that this coursework helped to develop feelings of competence. Such an increase in motivation would tend to increase participants' desire to develop skills in the future.

In addition, this course may also have led to increased singing self-efficacy and, possibly, general psychological well-being. Bandura (1977) described four major components to self-efficacy: mastery experiences, social modeling, social persuasion, and psychological responses. This coursework provided ample opportunity for mastery experiences, as participants practiced their skills repeatedly over time. They also were encouraged by their classmates and instructor (social persuasion), and they observed their classmates, which provided opportunities for social modeling. Self-efficacy is often considered to be domain-specific—that is, it is an indication of confidence in a particular area, i.e., singing. There is some research, however, that seems to support a Cross-Domain Influence Model (Jungert et al., 2014). Therefore, self-efficacy in one domain may lead to increased self-efficacy in another domain. In addition, other researchers have found indications of a bottom-up effect in which domain-specific self-efficacy might lead to overall self-efficacy (Grether et al., 2018). Increased overall self-efficacy has been linked to increased psychological well-being (Siddiqui, 2015). Although we did not measure any of the components of self-efficacy theory or self-determination theory, there is a possibility that the increases in confidence could be an indicator of increased psychological well-being. Future researchers might directly measure these components to determine whether there is a connection to overall psychological well-being.

Certainly, the small number of participants is a limitation in this study, which is exacerbated by the fact that not all 59 participants answered the open-response questions. Since this study included those who had taken the courses between fall 2014 and spring 2018, a follow-up study that includes students who have taken the courses between fall 2018 to the present would increase the number of subjects considered.

An additional limitation is that the students who answered the survey all took the same singing courses from the same instructor. Although there were small changes to the content that was delivered in the classes as they were taught over the years, they all followed what was essentially the same curriculum. Therefore, it should not necessarily be assumed that students who took this singing class would have the same experience as students who take a different singing class, or even the same class taught by a different instructor. Future investigators may seek participation from those in other college or university programs with a similar course that follows a similar format.

In the same vein, students were asked in the survey how the course impacted their confidence in auditions and in productions. They were not asked if improved confidence in singing made them feel more confident in their everyday life. Future studies may specifically investigate links between confidence in singing and confidence in other activities.

Conclusion

In this study, we focused specifically on singing within the context of a university theatre degree program, the results of which point to the positive impacts such training has on confidence. Given what was discussed above regarding self-efficacy, it is reasonable to believe that boosting confidence in the ability to sing in front of others may have an additional positive impact on overall confidence, as well.

Links between participation in the performing arts—whether music, singing, or theatre—and increased health and well-being are abundant in the research literature. The fact that this study aligns with this body of work is perhaps unsurprising. It is likely even less surprising among performing arts educators who are flooded with day-to-day examples and anecdotal evidence of

https://doi.org/10.52202/078697-0005

just how much individuals benefit in a variety of ways from engaging in the arts. Hopefully, this study provides even further justification for the vital work these practitioners are doing.

References

Aldridge, D., Schmid, W., Kaeder, M., Schmidt, C., & Ostermann, T. (2005). Functionality or aesthetics? A pilot study of music therapy in the treatment of multiple sclerosis patients. *Complementary therapies in medicine, 13*(1), 25-33.

Bandura, A. (1977). Self-efficacy: Toward a unifying theory of behavioral change. *Psychological Review, 84*(2), 191-215.

Borgwardt, L. (2010). Inspiring Creativity and Self-Confidence: The Power of Theatre in a Child's Life.

Clendenon-Wallen, J. (1991). The use of music therapy to influence the self-confidence and self-esteem of adolescents who are sexually abused. *Music therapy perspectives, 9*(1), 73-81.

Côté, I., Getty, L., & Gaulin, R. (2011). Aphasic theatre or theatre boosting self-esteem. *International Journal on Disability and Human Development, 10*(1), 11-15. doi: org/10.1515/ijdhd.2011.007

Demorest, S. (2018, March 1). *Stop obsessing over talent – everyone can sing.* The Conversation. https://theconversation.com/stop-obsessing-over-talent-everyone-can-sing-74047

Dweck, C. (2017). *Mindset: Changing the way you think to fulfil your potential* (updated edition). Little, Brown Book Group.

Fancourt, D., Finn, S., Warran, K., & Wiseman, T. (2022). Group singing in bereavement: effects on mental health, self-efficacy, self-esteem and well-being. *BMJ Supportive & Palliative Care, 12*(e4), e607-e615.

Grether, T., Sowislo, J. F., & Wiese, B. S. (2018). Top-down or bottom up? Prospective relations between general and domain-specific self-efficacy beliefs during a work-family transition. *Personality and Individual Differences, 121*, 131.139. doi: org/10.1016/j.paid.2017.09.021

Hillman, S. (2002). Participatory singing for older people: A perception of benefit. *Health education, 102*(4), 163-171.

Horne, C. A. (2017). *Developing Confidence in Late Adolescents: A New Theatre Curriculum* (Doctoral dissertation, University of Akron).

Jungert, T., Hesser, H., & Träff, U. (2014). Contrasting two models of academic self-efficacy – domain-specific versus cross-domain – in children receiving and not receiving special instruction in mathematics. *Scandinavian Journal of Psychology, 55*(5), 440-7. doi: 10.1111/sjop.12139

https://doi.org/10.52202/078697-0005

Lestari, R. K., & Handayani, S. S. D. (2017). Children Confidence Development by Theme Based Movement and Singing Method at RA Islamic Tunas Bangsa 4 Ngaliyan Semarang. *BELIA: Early Childhood Education Papers*, *6*(1), 14-17.

Manternach, B., & Manternach, J. N. (2019). Survey of University Acting Majors regarding the Most Useful Elements of Singing Training. *Voice and Speech Review*, *13*(3), 312-321.

McCammon, L. A., Saldaña, J., Hines, A., & Omasta, M. (2012). Lifelong impact: Adult perceptions of their high school speech and/or theatre participation. *Youth Theatre Journal*, *26*(1), 2-25.

Pavlicevic, M., O'Neil, N., Powell, H., Jones, O., & Sampathianaki, E. (2014). Making music, making friends: Long-term music therapy with young adults with severe learning disabilities. Journal of intellectual disabilities, *18*(1), 5-19.

Ryan, R. M., & Deci, E. L. (2000). Self-determination theory and the facilitation of intrinsic motivation, social development, and well-being. *American Psychologist, 55*(1), 68–78. https://doi.org/10.1037/0003-066X.55.1.68

Savatgy, M. (2022). Lights, Camera, Confidence: The Effects of Theatre on Self-Esteem.

Siddiqui, S. (2015). Impact of self-efficacy on psychological well-being among undergraduate students. *International Journal of Indian Psychology, 2*(3), 5-16.

Wu, S. M. (2002). Effects of music therapy on anxiety, depression and self-esteem of undergraduates. *Psychologia, 45*(2), 104-114.

Appendix One

List of phrases from the open-ended questions determined to be synonymous with "confidence"

1. To be relaxed and have fun!
2. Performing in front of my peers is invaluable, particularly for students like me who still are figuring their own voice out.
3. I think learning the fundamentals of anatomy and how singing is supposed to work helped me change my view on my own voice and be less ashamed of my own sound, encouraging me not to push and to let it just be what it is.
4. It helped singing in front of people less scary.
5. I think the most valuable thing I learned is to trust my own voice.
6. I think that as an actor it was important to realize that I can do some stuff with music without feeling completely out of my element. I think the practice of being up in front of my classmates and making acting choices while doing something I'm less comfortable with was useful to help me to realize that it's something I can do.
7. Learning how to sing in front of others and not be afraid.

https://doi.org/10.52202/078697-0005

Appendix Two

List of phrases from the open-ended questions determined to be synonymous with "skills"

1. The time I got to perform and practice the songs was helpful to develop my abilities.
2. I learned some of my tendencies while singing and I was given exercises to do to encourage relaxation while singing. I think learning the fundamentals of anatomy and how singing is supposed to work helped me change my view on my own voice...encouraging me not to push and to let it just be what it is.
3. The vocal pedagogy was incredibly important in actually making me sound better.
4. I also think having a good base for healthy singing technique and learning a few good vocal warmups to have in your back pocket are very important.
5. Provided a solid foundation for vocal improvement...vocal technique work has been a godsend in terms of helping me pinpoint areas to work on.
6. Helped me sing in a way that's healthier for my voice.
7. I loved when we got personal time with Brian to practice and work on individual improvements.
8. It helped me to build a stage presence as well as understand where I'm breathing from.
9. Learning vocal pedagogy, and general vocal health/good vocal use has helped me maintain my voice better than I've been able to before.
10. The rounding out and diversifying of oneself and one's capabilities as an actor is so valuable.
11. Vocal exercises helped in speech in acting.
12. I feel much stronger in my voice after having taken the class.
13. I think that being at least able to sing OK in an audition and performance setting is important, so I appreciated that this was part of the curriculum.
14. It really helped with my audition songs and technique.
15. I think oftentimes I think of acting and singing technique as two different things, but being able to integrate my singing technique into my acting technique while learning them both helped me find the connections between the two.
16. If my voice hadn't been ruined by cancer (now you know who I am) I would've had the confidence and ability to audition for musicals.

https://doi.org/10.52202/078697-0005

Exploring the Impact of Binaural Beats and Various Music Genres on Cognitive Performance and Concentration: A Pilot Study

Ana Paul

Abstract

The intention of this study is to explore the influence of binaural beats, in comparison to other sonic environments, on cognitive performance. The study participants were ten University at Buffalo students aged 18-22 (five men and five women) with no prior exposure to binaural beats. This was a within-subjects design involving three sonic environments: binaural beats, music of their choice, and silence. At first, the participants performed three cognitive training games in silence: Train of Thought, Color Match, and The Chalkboard Challenge, all of which are games found on Lumosity (2024), an online platform which aims to improve memory, attention, and problem solving through its games. A week later they were randomly divided into two groups. Group 1 performed the same tasks while listening to binaural beats. Group 2 performed the same tasks as the week prior, but they were allowed to listen to their own choice of music, which could be anything *except for* binaural beats. Different scores for the cognitive performance games were yielded across the sonic environments. Comparisons were made within each participant to determine which sonic environment yielded higher scores.

Keywords: Music & Focus, Music & Attention, Music & Memory

 https://doi.org/10.52202/078697-0006

Introduction

A binaural beat is an auditory illusion perceived by the brain. Exposure to 415 Hz in the left ear and 405 Hz in the right ear results in hearing a third tone that is not actually there. The result is a binaural beat with a frequency of 10 Hz (Colzato et al., 2017). When listening to an audio clip with a binaural beat of 10 Hz, the brain enters an alpha brainwave state, since 10 Hz corresponds to Alpha Waves. Subsequently, one enters a relaxed, focused state. Studies show that binaural beats corresponding to Alpha waves can have a temporary positive effect on working memory (Kraus & Porubanová, 2015) and can bring an increase of performance in cognitive processing (Cruceanu & Rotarescu, 2013). It is important to note that different binaural beat frequencies are related to different brain waves, thus they would be useful for various cognitive functions.

In this experiment I focused on 35 Hz and above, which in literature are associated with gamma waves, resulting in enhancing cerebral functions such as perception, attention, memory, consciousness, and motor control (Amo et al., 2017). I used binaural beats soundtracks made available to the public (Magnetic Minds, 2017) claiming to improve focus, with the aim of exploring the relationship between binaural beats and cognitive performance.

Method

This is a repeated measures experimental research design, though since the sample size is small, this could be considered a pilot study. Ten University at Buffalo students volunteered to participate in the pilot study ages 18-22, five men, and five women. The students had no prior knowledge of binaural beats. A short pre-study survey was completed, where students were asked if they prefer to study while listening to music or in silence. Their response was 41.2% reported that it's easier to study in silence while 58.8% responded that it's easier to focus while listening to music. I concluded that some students listen to music while studying and some do not. During the experiment, I aimed to determine if cognitive performance is optimal while listening to their music of choice, binaural beats, or while working in silence. The hypothesis was that listening to binaural beats improves scores on the tests.

The subjects performed three cognitive training games in silence: Train of Thought, Color Match, and The Chalkboard Challenge. A week later they were randomly divided into two groups. Group 1 performed the same tasks while listening to binaural beats this time. Group 2 also

https://doi.org/10.52202/078697-0006

performed the same tasks, but they were allowed to listen to their own choice of music, which could be anything *except* binaural beats.

Train of Thought is a test of divided attention. Divided attention refers to the ability to simultaneously respond to multiple tasks or task demands. The task is to guide an increasing number of trains to their stations using computer software. The subject must guide them all simultaneously at their stations.

For Color Match, the Stroop Test was administered. The Stroop Test challenges a person's selective attention. The test is to see how quickly one can identify if the text color matches the meaning of the text. For example, you will have the meaning written on the left, "red" and on the right you will have the word "yellow" written out with a red color. The task is to realize that the meaning "red" matches the color red while ignoring the fact that it says "yellow," and just look at the color.

The Chalkboard Challenge improves quantitative reasoning and problem-solving skills. Part of quantitative reasoning is numerical estimation — the ability to approximate numerical relationships quickly. One is given two mathematical equations and must quickly determine which one is larger. The first round of scores from group 1 were compared to the second round of scores from group 1, to see if there was any change in scores. Subsequently, the first round of scores from group 2 were compared with the second round of scores from group 2. After this I took note of which group had a higher average improvement rate.

Results

The Lumosity (2024) website provided the calculated scores for each of the tests. In order to measure the data, I took the scores after the tasks had been performed in silence, and I took the scores after the tasks were performed while listening to binaural beats or music of choice. The number of participants that improved their scores were averaged out and compared for both groups. The group listening to binaural beats yielded a 73.3% improvement rate and the group listening to the music genre of their choice yielded a 46.6% improvement rate.

https://doi.org/10.52202/078697-0006

Train of Thought

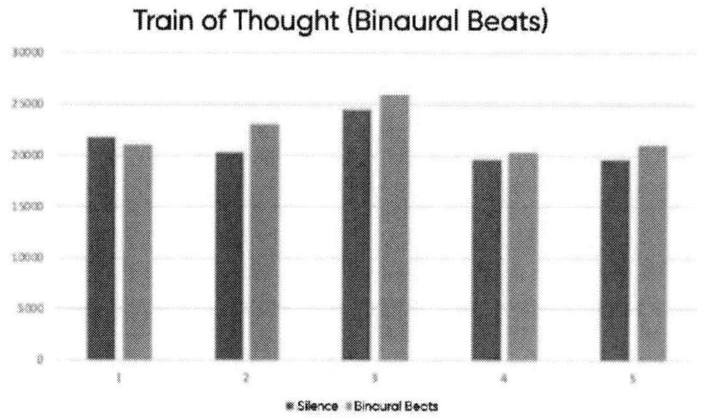

80% of students improved their scores when listening to binaural beats.

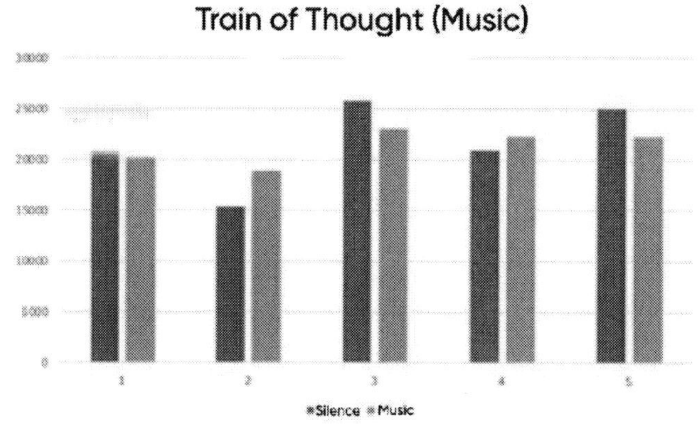

40% of the students improved their scores when listening to their own music choice.

Color Match (Stroop Test)

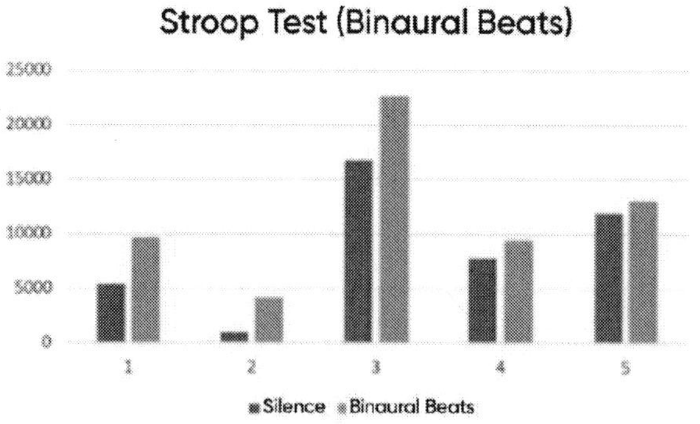

100% of students improved their scores when listening to binaural beats

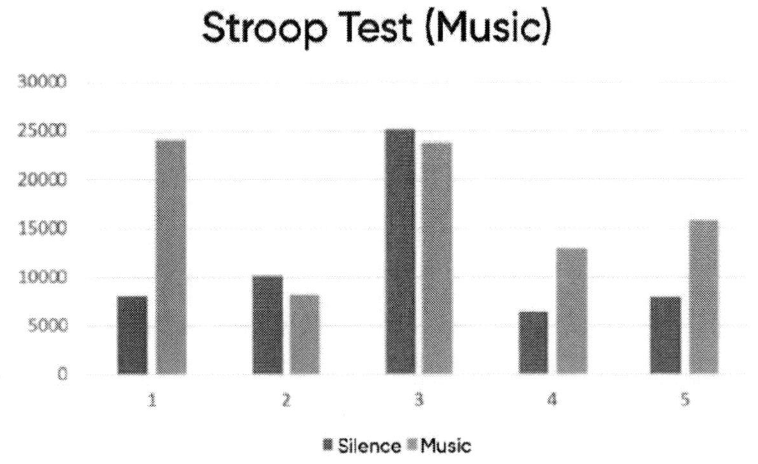

60% of students improved their scores when listening to their own choice of music

Chalkboard Challenge

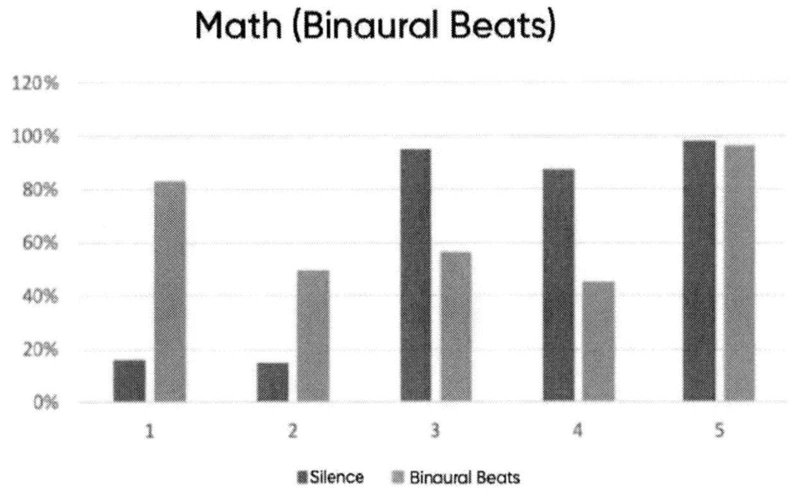

40% of students improved their scores when listening to binaural beats

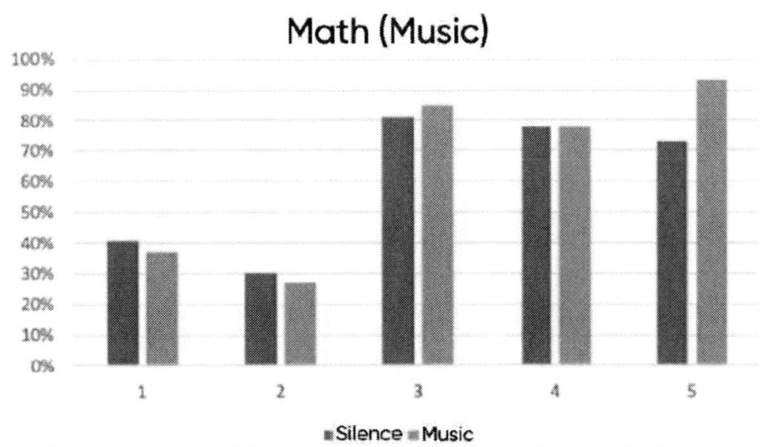

40% of students increased their scores when listening to their own music choice.

https://doi.org/10.52202/078697-0006

Discussion

From this pilot study I concluded that listening to binaural beats improves scores on cognitive training games more than music of choice. The method could have been improved by having participants listen to binaural beats for about ten to fifteen minutes before the start of the task in order to ensure that the brain has had the time to process the binaural beats, and that the participant was already in the focused state before starting the task. When this study is repeated, the binaural beats should also be less distinguishable, and played simultaneously with a recording of rain or the ocean waves as was done in a previous research study (Kraus & Porubanová, 2015), in order to ensure that the meditative music itself doesn't affect participant's brains. Participants would then be allowed to play their music of choice again. A song would be played twice, once with binaural beats in the background and once without.

It is also important to note that the research was only conducted with students at the University at Buffalo. The study may yield different results if performed in a non-academic setting and on people of different schooling backgrounds. Practice effects were a concern during this research study as well, because they could compromise the internal validity of the study, as participants could have improved their scores simply based off of doing the task twice. In an effort to avoid this, each student was administered the tests a week apart. Additionally, Lumosity (2024) generates slightly different tests each time, though the concept remains the same. It is also important to note that since we didn't have access to an EEG, we could not assess in which brainwave state the participants were in while they were listening to the binaural beats, so we cannot determine if the participants were indeed in a gamma state. A larger study is needed to confirm the findings of this pilot study.

https://doi.org/10.52202/078697-0006

References

Amo, C., de Santiago, L., Barea, R., López-Dorado, A., & Boquete, L. (2017). Analysis of gamma-band activity from human EEG using empirical mode decomposition, *Sensors, 17*(5), 989. doi:10.3390/s17050989

Colzato, L. S., Barone, H., Sellaro, R., & Hommel, B. (2017). More attentional focusing through binaural beats: Evidence from the global–local task. *Psychological Research, 81*, 271–277. doi:10.1007/s00426-015-0727-0

Cruceanu, V. D., & Rotarescu, V. S. (2013). Alpha brainwave entrainment as a cognitive performance activator, *Cognition, Brain, Behavior, 17*(3), 249–261.

Kraus, J., & Porubanová, M. (2015) The effect of binaural beats on working memory capacity, Studia Psychologica, *57*(2). doi:10.21909/sp.2015.02.689

Lumosity. (2024). *Chalkboard Challenge,* Lumos Labs, Inc. Retrieved from www.lumosity.com/brain-games/chalkboard-challenge-2

Magnetic Minds. (2017). *Magnetic Minds Meditations.* Spotify. Retrieved from https://open.spotify.com/artist/38CxTZztQ6gwUCxe9dEfOH

 https://doi.org/10.52202/078697-0006

Sound Therapy and Music Medicine: Biological Mechanisms

John Stuart Reid

Abstract

Many books have been written on the subject of "Sound Healing," a term which infers that sound has the ability to heal, when in fact only the body can heal itself. However, books on Sound Healing rarely discuss the biological mechanisms stimulated by sound frequencies or music. Also, although Music Therapy has been practised in clinical settings since the 1950s, with many books and scientific articles available on this subject, there is a dearth of literature focused on the related subject of Music Medicine, a new term which is now being used in clinical settings, relating to the beneficial effects of music for a patient, without the intervention of a Music Therapist.

The present article provides a short introduction to Sound Therapy and Music Medicine, including examining the biological mechanisms. An historical overview is presented first, in which evidence is discussed for the ancient Egyptians employment of an aspect of Quantum Medicine, although their understanding of imbibing water poured over a healing statue was based on magic, today we understand that placebo mechanisms can trigger the body's healing response through the power of belief. Aspects of the Greek philosopher Pythagoras' belief in "music as medicine" are discussed, before the article shifts to an overview of the origin of inaudible therapeutic ultrasound and audible sound therapy.

The main section of the article begins with definitions of Music Medicine and Sound Therapy, followed by discussions of chronic pain mediation by sound, musical stimulation of the immune system, sonic stimulation of the vagus nerve system, and several other sections that describe the application of sound frequencies and music to the full body, as distinct from the mechanisms stimulated by receiving sound via headphones.

Keywords: Sound Therapy, Music Medicine, Healing

https://doi.org/10.52202/078697-0007

Introduction

The intent of this article is to provide clinicians and other medical professionals with an introduction to the related fields of sound therapy and music medicine, and to outline some of the many biological mechanisms advantageously activated by sound and music, collectively categorised as "vibrational medicine." The information offered is intended to educate clinicians, nurses, medical students, integrative medicine practitioners and hospital and clinic administrators, in the therapeutic potential of vibrational medicine.

Since the development of quantum physics in the twentieth century, discoveries made in medical physics reveal the body as a complex interplay of biofields (Rubik, 2002) in which energy-information flows throughout the organism. At the level of the cell, information is exchanged through electromagnetic signals—primarily in the far infra-red spectrum—in addition to biochemical signals and sonic frequencies (Reguera, 2011). At the atomic level, biological complexities, and energy-information flow, can be viewed in terms of vibration. Nobel Laureate, Max Planck (1944), said:

> As a man who has devoted his whole life to the most clear-headed science, to the study of matter, I can tell you as a result of my research about atoms this much: There is no matter as such. All matter originates only by virtue of a force *which brings the particle of an atom to vibration* and holds this minute solar system of the atom together (Speech).

It is in this context that vibrational medicine has its roots: considering the energetic (vibrational) interconnectedness of the mind-body system. Practitioners of holistic medicine, or functional medicine (The Institute for Functional Medicine, 2024), as it is often referred to, review all aspects of the patient, including their emotions. In this expanded medical model, since the body is comprised of vibrational energy, a wide variety of vibrational and energetic modalities are available to support the patient's physiology, including sound and music.

Some of the physiological mechanisms initiated by sound therapy and music medicine are achieved by whole-body immersion in specific sound frequencies, or in music, either recorded or live. Other mechanisms, initiated neurologically, can be achieved by listening to specific sounds or music via headphones.

https://doi.org/10.52202/078697-0007

An important yet little discussed aspect of physics, with significant implications for medical science, is that all sounds, whether single frequencies or a complex array of musical frequencies, create near infrared light (NIR) and far infrared light (FIR), due to the atomic physics of inelastic sonic collisions. The infrared light created by sound and music is why sound intensity is measured in watts per metre squared (College Sidekick, 2024) and such light is modulated in amplitude by the sound, thus, carrying the FIR component of the sonic energy-information almost 4cm into the body's tissues (Vatansever & Hamblin, 2012) and in the case of NIR, up to 3 cm through the skull into brain tissue (Henderson & Morries, 2015). As intercellular communication occurs mainly in the far infrared spectrum, the physics of sound-light interactions infers that sonic-modulated light is conveyed to cells in the medium of their own "language" (Reguera, 2011).

The author's studies in acoustic-physics have revealed the quasi-holographic nature of sound in which the identical energy-information is contained throughout the sound "wave," whether emitting from a transducer, a musical instrument, or the human voice. However, the actual space-form of audible sound is spherical, therefore, the energy information conveyed by sound and music can be considered to exist at all points on the surface of the sound bubble and within its interior (Dennis, McNair, & Kauffman, 2013).

Depiction of a sound bubble emitting from a violin.
Audible sounds are quasi-holographic in nature and emit spherically from a sound source.
All points on the bubble's surface and within its interior, carry identical energy-information.
(Graphic by Dean Baker)

https://doi.org/10.52202/078697-0007

This important fact has far-reaching implications for the future of sound therapy and music medicine, particularly in relation to Faraday Wave patterns (also popularly known as "cymatic" patterns) that manifest on the surface membranes of cells, organs, fascia, and in visceral fluids.

The idea that sound, in the form of music, can be utilised as a therapy in support of illness is a concept espoused by Pythagoras of Samos, circa 2,500 BCE. One of his biographers, Iamblichus, reported that Pythagoras believed that music could be used in place of medicine and that it contributed greatly to health (trans. Thomas, trans. 1818). Even though his belief is an over-simplification, in fact, sound and music have been shown in recent decades to have many therapeutic effects, as the content of this article demonstrates. According to Socrates (436–338 BC), and other writers, Pythagoras of Samos (circa 570 – circa 495 BC), travelled in Egypt (Cantor, 1897; Joost-Gaugier, 2007; Stanley, 1687/2010) and absorbed knowledge from their priest-scientists. Music healers practised their craft in ancient Egypt and enjoyed privileged relationships with priests and high-ranking government servants. Music was called "physic for the soul" by Egyptian priest-physicians (Thaut, 2015).

The Egyptians built Sanatoria (hospitals) in all major towns. The foundations of the Dendera Sanatorium feature small healing chambers that would have had excellent reverberative qualities, due to their parallel facing walls and flat surfaces, with acoustics similar to modern bathrooms. The Egyptian music healers used conventional musical instruments such as the drum, harp, flute, lyre and tambourine, but 'sistra,' were also employed. The sistra is a type of rattle with metal discs that emits significant levels of ultrasound in the 40 to 60 kHz range, which could have supplemented the patients' beneficial results. At the Festival of Opet the sistra was used to stimulate the nostrils: *"Receive the sistra presented to your nostril that he may give rejuvenating breath..."* a statement suggesting that the ancient Egyptians were aware that the sistra emitted a specific quality of sound that caused a rejuvenating effect on the sinus cavities (Manniche, 1991). Each healing chamber featured a basin containing holy water from the adjacent sacred lake, the water's miraculous powers were amplified by pouring it over healing statues inscribed with appropriate texts (Nunn, 1997). This principle is not unlike the

https://doi.org/10.52202/078697-0007

modern placebo effect, which holds imagined power due to the influence of the patient's beliefs on their physiology, articulated by Lipton (2005) in *The Biology of Belief*.

Dendera Sanatorium

There is also some evidence that the ancient Egyptians employed vowel sound chant (as distinct from singing) for therapeutic effect, just as today humming and chanting are known to stimulate nitric oxide production and to stimulate the vagus nerve system, for example, therapies that will be discussed later in the article. The mere sounds of words were considered a potent and creative force in Egypt, as noted by Schwaller de Lubicz, (1982) in *Sacred Science*. Several textual indicators are known regarding the Egyptian's use of vowel sounds, of which the following are two examples.

In a quote from Aristotle in *Poetics*, a Greek traveller, Demetrius, circa 200 BCE, is reported to have said:

> In Egypt, when priests sing hymns to the gods, they sing the seven vowels in due succession, and the sound of these vowels has such euphony that men listen to it instead of the flute and lyre (p.395).

Such euphony would have been enhanced by vowel sound chant in chambers designed to be reverberative, an acoustic quality that is true of almost all ancient Egyptian temples and tombs.

A letter from Asclepius to King Ammon, from *The Corpus Hermeticum*:

> The very quality of the sound and pronunciation of the Egyptian language carries itself the power of what is being spoken...We do not use a noise of words [referring to the Greeks] but sounds full of power (trans. Salaman, Oyen, & Warton, 1999, p. 74, W.D. Gerald Duckworth & Co. Ltd).

Imhotep (2667-2600 BCE) was the most revered of all physicians in ancient Egypt and was deified by the Egyptians circa 525 BCE. During his lifetime he was vizier to King Djoser and architect of Djoser's Step Pyramid complex at Saqqara. His legendary abilities include poet, mathematician, astronomer, architect and physician, and in his medical treatise he regarded disease and injury as naturally occurring, not punishments sent by gods (Mark, 2016).

In the Ptolemaic Period, the Greeks identified Imhotep with Asklepios, their god of medicine, and in the second century BCE Ptolemy VIII built a shrine to Imhotep at the great temple of Hatshepsut at Deir el-Bahri, which then became a place of pilgrimage by the sick. The great physician, Sir William Osler pronounced in 1923 that Imhotep was the first figure of a physician to stand out clearly from the mists of antiquity (Thaut, 2015).

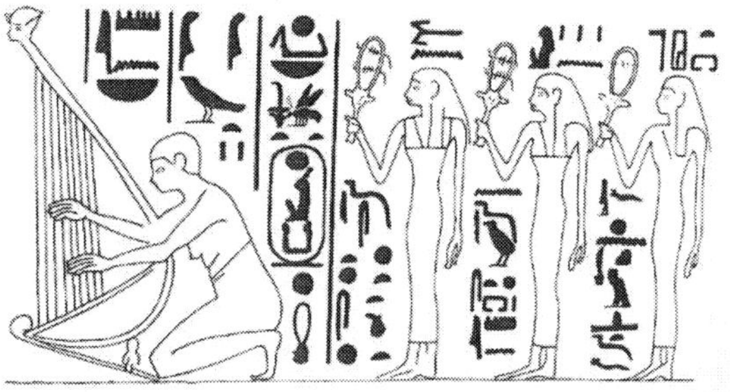

The Feast of Opet

In the scene from The Feast of Opet, in a building erected by Queen Hatshepsut at Deir el-Bahri, a male harpist and three female musicians perform with sistra instruments (Manniche, 1991, p. 72).

Returning to Pythagoras, he proposed that by listening to music one could comprehend and retrace the outer "physical" harmony of the universe, which would lead to a state of inner "mental" harmony, thus re-establishing balance in the body and helping to cure mental disorders (Thaut, 2015). He later founded an intellectual community on the island of Croton, Italy. One group of his followers were known as the "Acoustici", which means "to hear," and from which the word "acoustic" derives. While Pythagoras is credited with the invention of the

https://doi.org/10.52202/078697-0007

monochord—a single gut string tensioned by a fixed weight—which he used for investigating harmonic intervals, the ancient Egyptians had been making and playing musical instruments since pre-dynastic times, a history that had already spanned three millennia by the time of Pythagoras' visit. Therefore, it seems likely that Pythagoras would have absorbed knowledge of musical intervals, in addition to music as medicine, during his Egyptian travels.

Aristotle (384 BCE - 322 BCE), a pupil of Plato, viewed the therapeutic aspect of music as creating an emotional catharsis in mental states and spoke about music imitating passions or states of the soul (Thaut, 2015). An interesting account of music therapy with mental illness comes from Asclepiades of Bythnia (124–40 BCE), the first physician in history to create a health and disease theory resembling what is known today as molecular medicine. Asclepiades was a pioneer in the humane treatment of patients with mental disorders, moving patients out of dark confinements and prescribing work therapy, healthy diet, massages, and music therapy (Thaut).

Following Greek culture, no radically new theories and practises of music therapy were reported during a period of roughly 1500 years. However, in the seventeenth and eighteenth centuries there was a flourishing of writings on music and medicine, notably that of the German, Athanasius Kircher (1602-1680) who created an entirely new theory of music's therapeutic effects. He gave one of the most comprehensive descriptions of music in medicine in his work *Phonurgia Nova*, in which he laid out the foundation for a therapeutic music style that he called Iatromusic, which is based on acoustic vibrations. The music sets air into vibration, which sets the "corpuscles" of the body into vibration. Kircher proposed a body-mind directed connection: mechanical vibratory energy from the music creates a physical-physiological response in the body, which also brings psychological mind processes into healing harmony (Thaut, 2015).

The Origin of Inaudible Therapeutic Ultrasound and Audible Sound Therapies

Therapeutic ultrasound was discovered in 1927 by Professor R. Wood and his assistant, Loomis (Fyfe & Bullock, 1985) and was explored as a medical modality in 1938 by Raiman Pohlman at Charité University Hospital, Berlin. Pohlman demonstrated the therapeutic effects of ultrasonic waves in human tissues and went on to introduce ultrasonic physiotherapy as a routine procedure in medical practice (Mason, 2011).

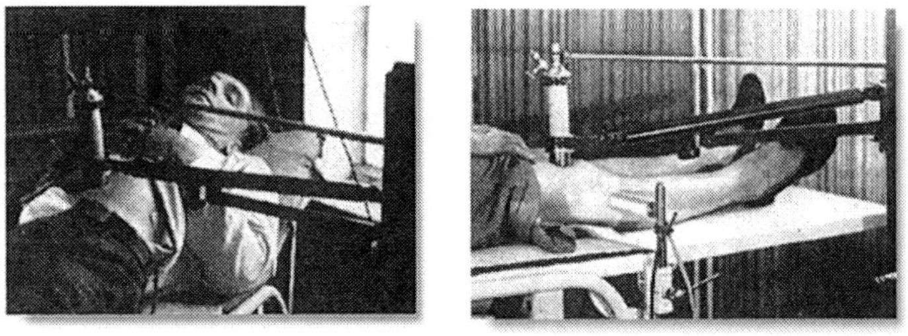

Raiman Pohlman, early therapeutic ultrasound apparatus
Left: Treating a gastric ulcer. Right: Treating arthritis.
(From *Les Ultra-sons appliques a la medecine*. Andre Denier, 1951)

Today, therapeutic ultrasound is typically applied at 1MHz, 2 MHz or 3MHz from a piezo-electric transducer, driven by an electronic oscillator and amplifier. Penetration depths of up to 5cm are achievable at 1MHz to support major physical trauma, while 2MHz and 3MHz frequencies are typically used to treat superficial conditions (Prentice, 1998). The therapeutic benefits of ultrasound are classified into thermal and non-thermal effects. With the discovery of ultrasound's medical properties, research burgeoned and it is now established fact that ultrasound has effective medical properties: breaking up kidney stones via ultrasonic lithotripsy (Marberger et al., 1985), shrinking tumours via high intensity focussed ultrasound (HIFU), histotripsy (Dubinsky et al., 2020), targeted drug delivery (Tian et al., 2020), and reducing brain plaque (Nicodemus et al., 2019) are a few examples of its many clinical applications. In hospitals and sports injury clinics throughout the world, therapeutic ultrasound is also used to support or accelerate the healing of soft tissues damaged by physical trauma, although the underlying biological healing mechanisms of therapeutic ultrasound remain poorly understood (Electrotherapy on the Web, 2024).

In 1928, Professor Erwin Schliephake, a German scientist who began his career at the Physiological-Chemical Institute in Leipzig, discovered that audible sounds create therapeutic

https://doi.org/10.52202/078697-0007

effects. His research inspired Mr. Lindacher, a colleague, to develop a hand-held, self-applied device that emits infra sonic vibrations (Novafon, 2022).

Professor Erwin Schliephake, and his infrasonic device

In the 1960s, English osteopath, Dr. Peter Guy Manners, developed an audible sound therapy modality that he named "Cymatic Therapy" that proved effective for the support of a wide range of ailments. ("Cymatics" is derived from the Greek word "kyma"—meaning "wave"—and was coined by Swiss medical doctor, Hans Jenny.) In recent years, several manufacturers have developed audible sound therapy devices (Cyma Technologies; KKT International; Medsonix), one of whom (Cyma Technologies) is continuing the work of Dr. Manners.

Music Therapy, Music Medicine, and Sound Therapy Definitions

Music Therapy is an accepted form of complimentary therapy in many hospitals and clinics, and may be defined as:

> The clinical and evidence-based use of music interventions to accomplish individualized goals within a therapeutic relationship by a credentialed professional who has completed an approved music therapy program. (American Music Therapy Association, n.d.)

Music Therapy is a proven modality but is limiting in the sense that each patient requires a music therapist with whom to work. A plethora of books and scholarly articles are available on the subject of Music Therapy, and therefore is not the focus of this article.

https://doi.org/10.52202/078697-0007

Music Medicine may be defined as:

> Listening to music [for the purpose of healing] without the presence of a therapist" (Bradt et al., 2015). Music Medicine is a relatively new clinical modality that refers to the therapeutic utilisation of music, chosen by the patient in a clinical setting without the intervention of a therapist. As its title implies, Music Medicine focuses on the demonstrable benefits of music as treatment for specific health challenges. The mechanisms by which music affects the body's systems are complex and this article provides a short introduction into this subject.

Sound Therapy is defined by the International Sound Therapy Association as:

> The application of audible sound to the full body or to a specific part of the body, from electronically-generated sound sources, or from musical sources, as therapeutic support, by a credentialed Sound Therapy practitioner (International Sound Therapy Association, 2024).

This definition clarifies that therapeutic audible sound can be generated by electronic means, or provided by a musical source. The biological mechanisms triggered by such sonic support will be discussed later in this section of the article.

Meta-analyses of Music Medicine in Clinical Trials

A Cochrane analysis of twenty-six music medicine clinical trials with a total of 1,369 participants titled, *Music for stress and anxiety reduction in coronary heart disease patients,* concluded that "listening to music may have a beneficial effect on systolic blood pressure and heart rate in people with coronary heart disease and appears to be effective in reducing anxiety in people with myocardial infarction." The same report mentioned, "Listening to music may reduce pain and respiratory rate and appears to improve patients' quality of sleep following a cardiac procedure or surgery" (Bradt et al., 2013).

A study by the University of Belgrade School of Medicine, led by Dr. Predag Mitrovic, involved 350 patients diagnosed with a heart attack and early post-infarction angina. Half were randomly assigned to receive standard treatment while half were assigned to regular music sessions in addition to the standard treatment. Researchers fine-tuned the music selection by working with the patient to determine the optimal music tempo and tonality, therefore, the

https://doi.org/10.52202/078697-0007

initial stage of the study would technically be classed as music therapy, while the major part of the study would, today, be classified as music medicine since the participants listened to music at home, without a therapist. Participants who listened to music reported less anxiety and less pain sensation than participants who received the standard treatment alone. Dr. Mitrovic concluded, "Sedative music was more effective than treatment in decreasing anxiety and pain in patients with HT [hypertension] and EPA [post-infarction angina]. Patients with HT should have the benefit of using sedative music as an adjuvant to medication during EPA episodes" (Napoli, 2020).

Johns Hopkins Medicine also acknowledges the role of music in addressing illness and indicates a range of illnesses they intend to treat with music, including, Huntington's Disease, Parkinson's Disease, and Dementia (John Hopkins Center for Music & Medicine, 2024).

McGill University in Montreal is also conducting studies in music medicine. In a meta-analysis of 400 studies, Dr. Daniel J. Levitin and Dr. Mona Lisa Chanda found that music improves the body's immune system function, reduces stress, and was found to be more effective than prescription drugs in reducing anxiety before surgery. They also found that listening to and playing music increases the body's production of the antibody immunoglobulin A and natural killer cells, boosting the immune system's effectiveness, as well as reducing levels of the stress hormone cortisol (McGill Newsroom Institutional Communications, 2013).

At Riuniti hospital in Ancona, Italy, neurosurgeon, Dr. Roberto Trignani performed an operation to remove a double tumour in the spinal cord of a ten-year-old boy, while molecular biologist and pianist, Emiliano Toso, played a grand piano in the operating theatre.

https://doi.org/10.52202/078697-0007

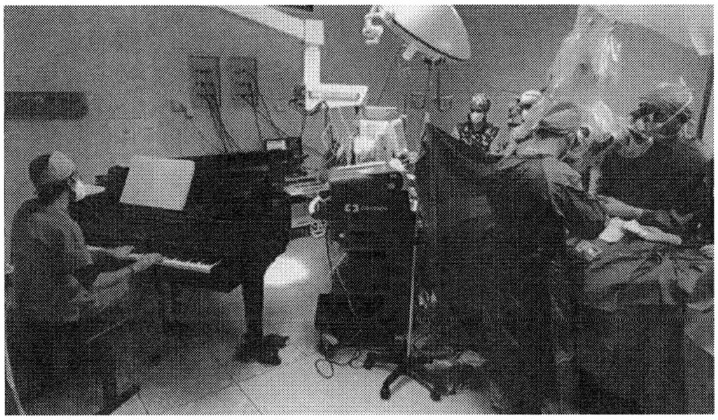

Dr. Emiliano Toso playing piano in an Operating Room,
during live neurosurgery

Monitoring the boy's brain activity via an encephalogram, suggested that the boy perceived the music. Dr. Toso said, "We tried stopping then restarting the music, noticing the patient's response. Despite the fact that the boy was under total anaesthesia, his brain appeared to perceive the music and this was very exciting." Dr. Trignani, head of the neurosurgery unit of Riuniti Hospital, commented, "Everything went well, there were no complications and there was a magical atmosphere of complete harmony in the Operating Room" (Fitz-Gibbon, 2020).

It is admirable and noble that musicians contribute their time and talent to playing in hospitals. The harp, in particular, has a long history of use in clinical settings and nursing homes and it is likely that it will always form an important aspect of patient care. However, to reach larger numbers of patients, and to give individual choice of music genre, a convenient solution is that of "modulated ultrasound speakers" (Holosonics.com, n.d.; Ultrasonic Audio Technologies, 2024), which emit wavelengths of only a few millimetres, thus, the acoustic energy travels as a narrow beam. Ultrasound contains frequencies far beyond the range of human hearing and is completely inaudible. However, as the low energy ultrasonic beam travels through the air, the inherent properties of the air molecules within the beam give rise to frequency components in the audible band that can be heard only by the patient to whom the beam is directed. In this future scenario, the patient would choose the music that they prefer from a bedside console, and be beamed to them from an ultrasonic ceiling-mounted speaker. Adjacent patients would not hear the music. In the illustration, a patient is listening to music

 https://doi.org/10.52202/078697-0007

without headphones, yet the nurse cannot hear the music.

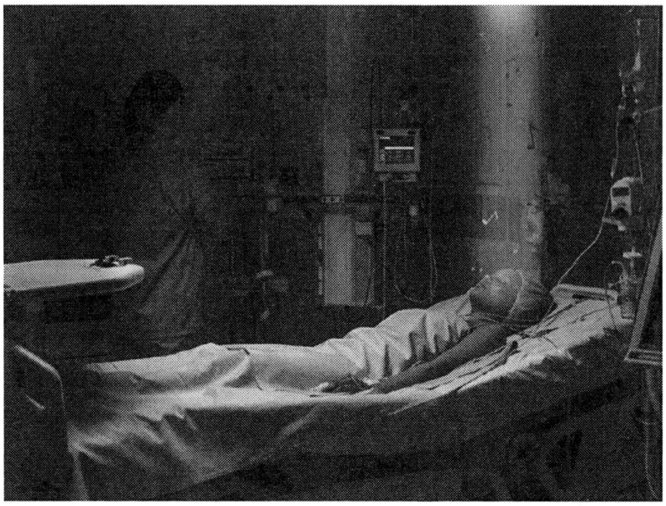

A patient listens to music without headphones and without disturbing the nurse or other patients

A brief overview of some of the biological mechanisms activated by full body immersion in music or in specific sound frequencies

(Detailed explanations are provided later in the article.)

Full body immersion in music or in specific sound frequencies (as distinct from listening with headphones), activates several beneficial biological mechanisms, five of which

are briefly summarised as follows:

- Enhances nitric oxide (NO) production through active and passive acoustic stimulation of the paranasal nasal cavities and lungs by specific sound frequencies and music, resulting in a broad range of health benefits.

- Promotes pain mediation through stimulation of the body's large A-beta fibres or A-alpha fibres in the area experiencing pain, thus causing the pain "gate" to close.

- Increases the availability of oxygen binding to haemoglobin molecules by low frequency sound pressure, thus breaking the pain-spasm-pain cycle or 'splinting cycle' by increasing the availability of oxygen to affected tissues.

https://doi.org/10.52202/078697-0007

- Activates the meridian system, via "sonopuncture," with many health benefits, including pain mediation.

- Increases the availability of oxygen binding to haemoglobin molecules by low frequency sound pressure, thus aiding tissue repair mechanisms.

A brief overview of some of the biological mechanisms activated by headphone listening to music, or listening to specific sound frequencies

(Detailed explanations are provided later in the article.)

- Mediates pain by the "Descending Inhibition of Pain" system, also referred to as the "top-down" modulation of pain. Such effects can be initiated by music (or white noise) as a result of activating endogenous opioids.

- Promotes stress reduction with consequent reduction in blood pressure and cortisol levels, and induces a state of joy with consequent increase in dopamine levels, leading to a proliferation of leukocytes, thus boosting immune system efficiency.

- Stimulates the brain binaurally—by binaural beats—to create changes in brain state, with physiological benefits.

- The vagus nerve system is stimulated, thus regulating internal organ functions, including digestion, heart rate and respiratory rate, as well as promoting vasomotor activity and anti-inflammatory effects. Specific very low (sub audible) frequencies may also be applied by full ear headphones, combined with music.

Each of these biological mechanisms will be discussed separately.

Active and passive sonic stimulation of the nasal cavities and lungs

Before the subject of sonic stimulation of the paranasal cavities and lungs is discussed, it is important to outline some of the natural health benefits of nitric oxide (NO), which is naturally produced in many areas of the body including the cilia in the nasal cavities and the alveoli in the lungs. NO reduces blood pressure by vasodilation (Cockroft, 2005) and many other health benefits are derived from this important molecule, for example: promotion of wound healing by cellular proliferation and angiogenesis (Rizk et al., 2004), mediation of cutaneous oedema

https://doi.org/10.52202/078697-0007

and inflammation, cytotoxic action against pathogens (Moilanen & Vapaatalo, 1995), increases cerebral blood flow and oxygenation to the brain, inhibits the aggregation of platelets within blood vessels thus helping prevent thrombotic events (Loscalzo, 2001), supports reduction of pulmonary hypertension and chronic obstructive airway disease (Ashutosh et al., 2000).

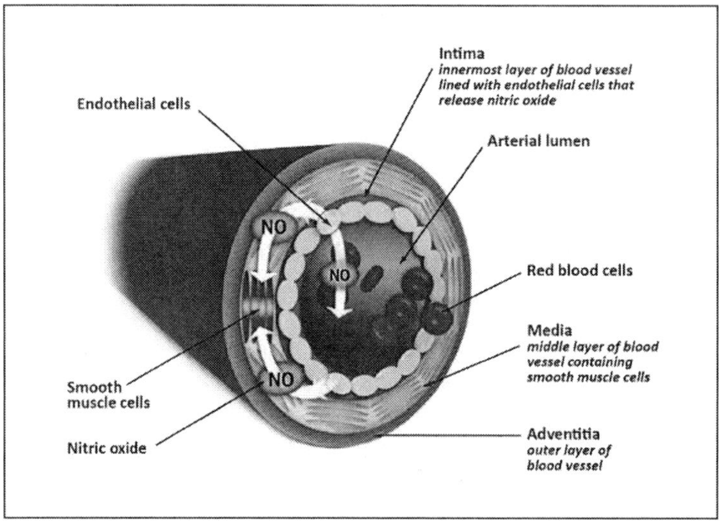

Nitric oxide relaxes smooth muscle cells in the walls of blood vessels, resulting in vasodilation
(Courtesy of NIOX)

NO can be produced in the body from the inorganic nitrates in green leafy vegetables and from fruits, particularly by the oral microbiome (Jones et al., 2021) and is also stimulated by exercise (Maiorana, O'Driscoll, Taylor, & Green, 2003) which can form part of a rehabilitation program, but the initial focus in this section is NO production in the paranasal cavities brought about by both active and passive sonic stimulation. 'Active' stimulation refers to the practise of vocal humming, which has been shown to greatly elevate NO production (Weitzberg & Lundberg, 2002; Goldman & Goldman, 2017). The movement of air across the nasal cilia generates NO, from which the many health benefits are derived, although the exact mechanisms by which NO is produced by the nasal cilia is not fully understood (Lundberg & Weitzberg, 1999).

The practise of nasal breathing is well known in the Yogic practise of pranayama, which means "breath control" in Sanskrit, a practise that is mentioned in the Bhagavad Gita, written at some point between 400BCE and 200BCE (Chetan, trans., n.d.).

In a paper titled, "Assessment of nasal and sinus nitric oxide output using single-breath humming" (Maniscalco et al., 2003), the authors show that NO is significantly increased by a single breath exhalation while humming, as shown in the graph.

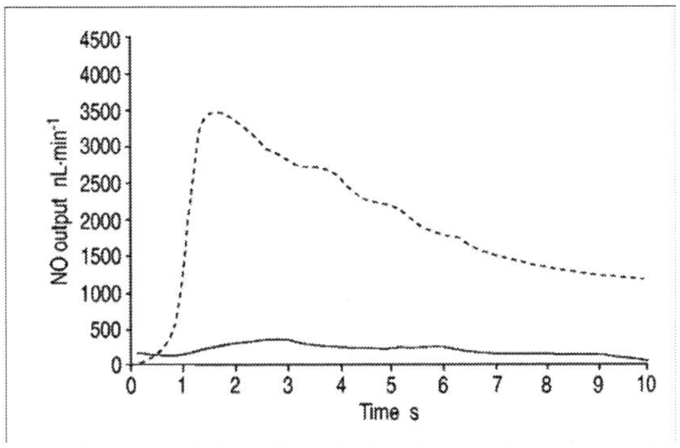

Original tracing of Nitric Oxide during a single-breath nasal exhalation with (dotted line) and without humming (solid line)

The authors of this study also carried out experiments to determine the optimum humming frequencies, and concluded that a measured frequency of 130Hz created the highest NO output of the sinus cavity in a human subject. The study does not specify whether the human subject was male or female but in either case the result is surprising when remembering that the sinuses consist of relatively small cavities, featuring Helmholtz resonant frequencies in the range 1kHz to 2kHz (Masuda, 1992) depending on gender and maturity.

Adult female skulls and sinus cavities are typically smaller than those of adult males; smaller sinus cavities support higher resonant frequencies. It should also be remembered that humming does not generate a single frequency but gives rise to an array of harmonics and that the prime resonant mode of the sinus cavities is automatically 'selected' during vocal humming as a natural aspect of Helmholtz resonance (the resonant property of a gas-filled cavity). Therefore, although the fundamental humming frequency of maximal excitation was found to be 130Hz, (Maniscalco et al., 2003), the sinus cavities would almost certainly have been excited by a specific harmonic of this frequency.

https://doi.org/10.52202/078697-0007

Nitric Oxide is also generated by the alveoli in the lungs (Pietropaoli et al., 1999) and can be stimulated by both active and passive sonic stimulation; actively by humming or singing, and passively by externally applied sonic frequencies or music. Indicators regarding the optimum frequencies for passive stimulation can be obtained from studies in which the respiratory system has been modelled in terms of its resonant sonic characteristics (Henry & Royston, 2017; Bogomolov et al., 2019). In the study by University of Illinois (Henry & Royston), the Helmholtz resonant frequency of a healthy volunteer is shown to be in the order of 100Hz, increasing to around 250Hz for a person suffering from pulmonary fibrosis. These frequencies will vary between individuals due to gender and lung capacity as a function of the patient's genetic makeup. Similarly, the Helmholtz resonant frequencies of the sinus cavities will vary between individuals.

Identifying the precise resonant frequencies of the patient's lungs or nasal cavities is not necessary to offer therapeutic intervention if the practitioner applies either broadband white noise (Britannica, 2024) or music to the patient. Both of these sound sources contain thousands of frequencies and Helmholtz resonance principles are such that a cavity will automatically 'choose' the frequency from a broadband sound source at which the cavity is naturally resonant. In the case of white noise, this sound source can be filtered through a parametric equaliser (Parametric Equaliser, 2024) that facilitates the selection of specific bands of frequencies that are appropriate for stimulating either the lungs or sinus cavities. By such filtering of white noise, any unpleasant sensations associated with the noise experience can be minimised. White noise generators are commercially available from many manufacturers and the audio signal from the generator drives a conventional audio amplifier and speaker. The speaker(s) should offer a frequency response that covers the range needed, e.g., ideally as low as 20Hz for excitation of lungs, and as high as 2kHz for excitation of sinus cavities. Commercially manufactured reference monitor speakers, which are typically used in recording studios, are ideal for this purpose because they offer an extended low frequency response and are typically fitted with 'tweeters' for reproduction of high frequencies.

These same principles in which a patient's sinus cavities or lungs automatically choose the specific frequency at which the cavity is naturally resonant, also apply to many musical instruments, for example, harps and gongs, whether played live in a clinical setting or via high fidelity reproduction equipment.

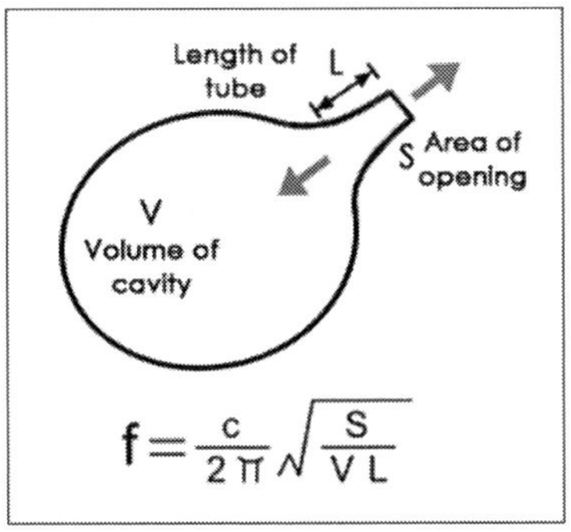

Helmholtz Resonance

f = resonant frequency of cavity, c = speed of sound in air, S = area of opening,
V = volume of air in the cavity, L = length of the tube

When applying recorded music to a patient for the purposes of lung and/or sinus activation, parametric filtering is not required because music provides a pleasant experience, assuming that the sound pressure levels are modest, e.g., 60 to 80dBA when measured in close proximity to the patient. The patient should make the music choice, although it is important that the music chosen contains a wide range of frequencies to ensure optimal stimulation of the lungs and sinus cavities.

In addition to stimulation of nitric oxide production, acoustic stimulation of the sinus cavities and lungs can also help to clear mucus and improves symptoms of Chronic Obstructive Pulmonary Disease (COPD) and chronic bronchitis (Sethi et al., 2014). In a 26-week randomised control trial the “Lung Flute” respiratory device, which produces a low frequency acoustic wave, the symptoms and health status of people with COPD and chronic bronchitis improved. The authors stated that mucus hypersecretion and impaired mucociliary clearance is prevalent in many patients with COPD, and contributes significantly to the morbidity and mortality of this disease, and concluded that the Lung Flute is a safe and effective treatment in COPD with chronic bronchitis (Sethi et al.).

https://doi.org/10.52202/078697-0007

Chronic pain mediation by audible sonic stimulation of nociceptors

Pain is a vital function of the body, providing early warning of damage or potential damage. It is both a sensory and emotional experience, affected by psychological factors such as past experiences, beliefs about pain, fear, or anxiety (Reddi et al., 2013). Tissue injury, for example, initiates the liberation of various inflammatory mediators, including prostaglandins, cytokines and chemokines. Leukocyte migration to the injured area, a characteristic of the inflammatory response, is associated with pain and tenderness, and is involved in wound healing (Wang et al., 2009). Acute pain is well understood and can be successfully mediated by analgesic medications; it is not within the scope of this article.

Chronic pain is a common, complex, and distressing problem, which has a significant impact on individuals and society (Mills et al., 2019). Chronic pain, like most diseases, often arises from a series or combination of multiple events (Mills, et al.). The biological processes that lead to the chronic pain state further increase sensitivity to painful stimuli and perceived levels of stress, which further modifies pain-related gene expression, creating a pathological pain cycle (Diatchenko et al., 2013). Even when there is a solitary precipitating event in the genesis of chronic pain (e.g. injury), there remain a series of factors that affect the duration, intensity, and effects (physical, psychological, social, and emotional) of chronic pain (Mills, et al.).

The International Association for the Study of Pain, define pain as "An unpleasant sensory and emotional experience associated with, or resembling that associated with, actual or potential damage" (Raja, et al., 2020) and chronic pain is "pain which has persisted beyond normal tissue healing time" (Bonica, 1953). Pain is regarded as chronic when it has lasted for more than three to six months (Merskey & Bogduk, 1994, p. 1).

Considering that pain is a universal experience, it is not understood why only a relatively small proportion of humans develop a chronic pain syndrome (Rodriguez-Raecke et al., 2009). Prolonged use of analgesics, such as chronic opioid therapy, is associated with constipation, sleep-disordered breathing, hypothalamic-pituitary-adrenal dysregulation, fractures (as a result of osteoporosis) and significant declines in health-related quality of life and increased health care costs Baldini et al., 2012). Therefore, it would be advantageous to mitigate chronic pain

https://doi.org/10.52202/078697-0007

without long term use of analgesics. In this section we discuss audible sonic stimulation of the body's nociceptors, as an alternative therapeutic modality in the treatment of chronic pain. Unlike prolonged use of analgesics, audible sonic interventions have no known adverse side effects.

Nerve signal conduction by sound

To lay a foundation for discussion in the principles of pain mediation by sound, it is important to mention discoveries concerning nerve signal transmission by sound.

In 1952, Alan Hodgkin and Andrew Huxley, working with a squid's giant axons, described how action potentials (or nerve impulses) in neurons are initiated and propagated, known today as the Hodgkin–Huxley model (Schwiening, 2012). It is regarded as one of the great achievements of twentieth century biophysics, for which they received the Nobel Prize in Medicine in 1963. Their theory, involving the flow of electric currents in nerves, became the standard teaching model in medical and biology textbooks. However, one aspect that puzzled researchers was the relatively slow conduction speeds in nerves, when compared with conduction speeds of electric currents in conductors. The speed of light in vacuum is 2.998×10^8 metres per second, which is approximately equal to a distance of 30cm per nano second. The speed of an electrical signal in a coaxial cable is about 2/3 of this, or 20cm per nano second, therefore, in one second the signal in a coaxial conductor will travel approximately 200,000,000 metres, which equates to a little over half the distance between the earth to the moon.

Nerve fibres, by comparison conduct signals several orders of magnitude slower than that of coaxial cables. The highest conduction speeds for nerve fibres are those of muscle axons, which can achieve speeds of over 100 metres per second.

The comparison with coaxial cables is made because in the four main classifications of primary afferent nerve fibres, three of them (A Alpha, A Beta, and A Delta) are myelinated, meaning that the nerve axon is enclosed in a myelin sheath, therefore resembling the dielectric material of a coaxial cable. Type C nerve fibre is unmyelinated, resembling a single electrical conductor. With coaxial cables, the speed of conduction is a function of the diameter of the fibre and the diameter of its outer sheath, and this is similar for myelinated nerve fibres; the larger fibres feature higher conduction speeds.

https://doi.org/10.52202/078697-0007

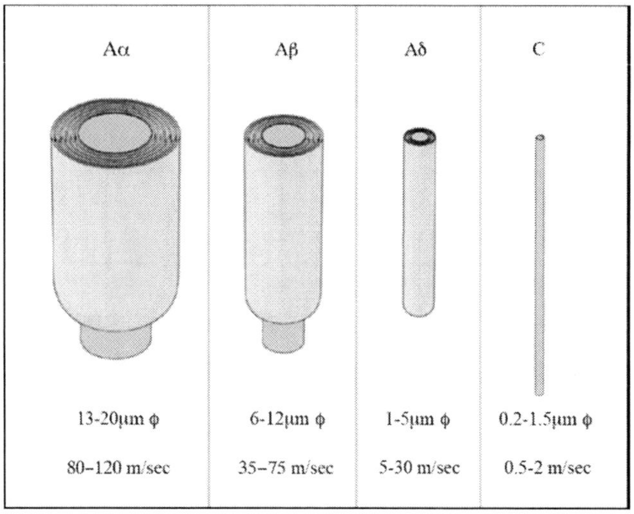

The main classifications of afferent nerve fibres and their conduction speeds, which are very slow compared with electric currents in coaxial cables

However, in 2005 a new model of nerve conduction was proposed by researchers at the Niels Bohr Institute at the University of Copenhagen, whose experiments showed that nerves conduct sound (soliton impulses), which in turn generate electrical pulses, due to the piezo electric effect (Heimburg & Jackson, 2005). In their paper they note that "...measured propagation velocities, which are ~100 m/s in myelinated nerves, find a satisfying explanation." Put differently, propagation of nerve impulses by sound explains the slow conduction speeds, while such sonic impulses give rise to electrical impulses that travel to the brain for interpretation.

Sonic propagation of nerve signals was first proposed by Konrad Kaufmann, in a 1989 unpublished paper (Kaufmann, 1989). Other researchers have found similar results, supporting the Neils Bohr Institute findings to a large degree (Shrivastava et al., 2018; Shrivastava & Schneider, 2014.). This discovery has significant implications for Sound Therapy and Music Medicine, particularly for whole body immersion in music and specific sound frequencies.

Principles of pain mediation by sound

Nociceptors are the specialised sensory receptors responsible for the detection of noxious (unpleasant) stimuli, transforming the stimuli into electrical signals, which are then conducted to the central nervous system (Reddi et al., 2013). They are the free nerve endings of primary afferent fibres and are distributed throughout the tissues of the body, including the skin,

https://doi.org/10.52202/078697-0007

viscera, muscles, joints and the meninges of the brain, (though not in the grey matter of the brain).

The four main classifications of afferent fibre have specialised roles, for example, response to light touch, or to acute events, or response to chemical or thermal stimuli, but crucially all types of afferent nerve fibre respond to mechanical pressure. And since sound may be defined as: "Mechanical radiant energy that is transmitted by longitudinal pressure waves in a material..." (Merriam-Webster, n.d.) it becomes clear that all types of afferent fibre respond to sound. This fact is reinforced by the Niels Bohr Institute discovery that nerves conduct sound as soliton pulses.

When nociceptors are stimulated, nerve impulses are transmitted to three spinal cord systems: the cells of the substantia gelatinosa in the dorsal horn; the dorsal-column fibers that project toward the brain; and the first central transmission (T) cells in the dorsal horn (Melzack & Wall, 1965).

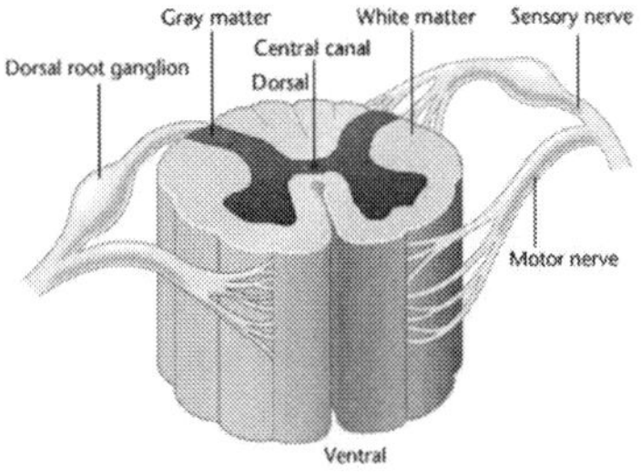

Cross section through spinal column showing the dorsal root ganglion
(Graphic: Emri Terim)

The theory by which sound frequencies can mediate pain are based on the "Gate Control Theory of Pain," which was first proposed in 1965 by Ronald Melzack and Patrick Wall (Malzack & Wall). The theory was initially met with scepticism, but despite having to undergo several modifications, its basic conception remains unchanged. Their theory provides a physiological-neural explanation for pain perception and ultimately revolutionized pain research. Gate Control Theory proposes that there are gates between the afferent nerves and the brain, located in the spinal column, which control how pain messages flow from the peripheral nervous system to the central nervous system.

 https://doi.org/10.52202/078697-0007

For example, pain-signals that conduct freely along small A-delta afferent fibres (which sense sharp pain) and small type C afferent fibres (which sense dull pain) open the gate, resulting in the perception of pain in the brain. By stimulating the large A-beta fibres or A-alpha fibres in the area experiencing pain, a reaction is caused in nearby inhibitory neurons. Once activated, Inhibitory Neurons, which sit on the same path as the Projection Neurons, the gate closes, thus muting pain signals before they reach the brain. Stimulating A-beta fibres or A-alpha fibres can be achieved by specific sound frequencies, as mentioned below.

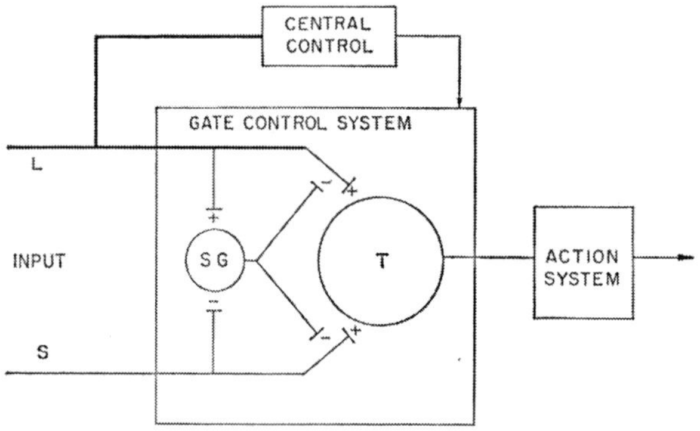

The Melzack-Wall Gate Control System (from Melzack and Wall)
L = large diameter nerve fibers, S = small diameter nerve fibers. The fibers project to the substantia gelatinosa (SG) and first central transmission (T) cells.
Activity in large fibers inhibits signals from small fibers.
(Drawing from Ronald Melzack & Patrick Wall: Pain Mechanisms: A New Theory.)

Some of the optimal frequencies found to be beneficial for pain mediation via nociceptor stimulation were discovered in Finland by clinical psychologist, Petri Lehikoinen, in the range 27Hz to 113Hz. Lehikoinen developed a therapeutic system: Physio Acoustic Sound (PAS) therapy, that was approved in the USA by the Federal Drug Administration (FDA) and in the UK by the British Standards Institute (BSI) for three claims: decreased pain, increased blood and lymphatic circulation and increased muscle relaxation and mobility (Bartel & Mosabbir, 2021). In Norway, Olav Skille placed particular emphasis on specific therapeutic frequencies at 40Hz, 52Hz, 68Hz and 86Hz (Bartel & Mosabbir).

Vibratory analgesia was also studied by Salter and Henry, examining P_1-purinergic receptors in the dorsal horn by adenosine, the neurotransmitter. In their study with cats, sound

https://doi.org/10.52202/078697-0007

at 80Hz induced depression of lower lumbar nociceptive neurons, which remained affected for up to four hours after cessation of the sound. The depression of the neurons revealed that adenosine was responsible for the analgesic effect, suggesting that gate control may be mediated by the release of adenosine resulting from the application of sound (Bartel & Mosabbir, 2021). This research shows that pain mediation by vibratory analgesia is not only active during focal application of low sound frequencies to the site of pain, but that the analgesia can be maintained for several hours after stopping the sound therapy due to the slow diffusion rate of adenosine (Sawknok, 2016).

Whole Body Vibration (WBV) is also becoming a popular therapy for low back pain and to improve functional abilities of patients, however the term "Whole Body Vibration" appears to be something of a misnomer in the sense that it has come to mean vibration applied specifically via the feet, that is, a vibrating platform of which the patient stands. However, WBV can be adapted to use in beds, as mentioned below. From seven randomized control trials involving 418 patients, four studies using pain as an outcome measure showed that WBV had a beneficial effect on pain compared with the control group (Wang, W., 2020). Although current WBV platforms can produce vibration frequencies up to 100Hz, frequencies below 30Hz are most commonly used (Bartel & Mosabbir, 2021).

Apart from their use in pain mediation and pain management, WBV deployed via a bed can support patients in an Intensive Care Unit (ICU). Critical illnesses can lead to a mean weight loss of 17% within the first 10 days of stay in the intensive care unit and up to 40% within 4–6 weeks. Early mobilization has a variety of beneficial effects, including mobility, reduced risk of venous thromboembolism, pneumonia, in addition to socioeconomic advantages. Effective active rehabilitation requires a conscious patient who is willing to participate, whereas WBV can be applied passively in bed (Boeselt et al., 2016).

https://doi.org/10.52202/078697-0007

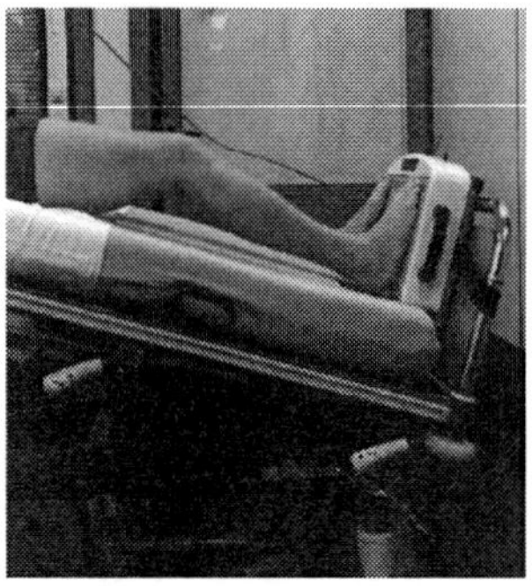

Whole Body Vibration

Deployed to the patient by fixing the device to the foot of the bed
(Courtesy of German Centre for Lung Research)

The WBV device was attached to the foot of the patient's bed, thus inducing vibrations into the bed frame. The patient lay in the supine position and the bed was inclined by 25° to transmit a sufficient load to the patient's lower limbs. Many beneficial health parameters were reported in the study but crucially, a significant increase in the patients' oxygen saturation level (SaO2) was seen in the second period, between the first and second minute (Mosabbir et al., 2020). This increase in oxygen binding to haemoglobin molecules by low frequency sound may have important implications for pain, as mentioned below, in the section: Breaking the 'pain-spasm-pain' cycle in spinal injury.

In a study titled, "The Effects of Long-Term 40Hz Physioacoustic Vibrations on Motor Impairments in Parkinson's Disease: A Double-Blinded Randomized Control Trial," the authors focussed on testing the efficacy of 40Hz vibrations in support of Parkinson's disease, a progressive neurodegenerative disorder. Symptoms are typically characterised by tremor, rigidity, bradykinesia and postural instability (Mosabbir et al., 2020). 40Hz was chosen because among the range of gamma frequencies (25-140Hz) it has shown consistent evidence for neuroprotection. The physioacoustic therapy employed was that of special reclining chairs with built in low frequency transducers. The 40Hz audio signal was inputted to the chairs using a scanning technique that caused the frequency to vary between 39.96Hz and 40.06Hz, resulting in a kinaesthetic pulsating effect. The placebo control group also sat in a reclining chair band heard a 40Hz tone without receiving any bodily vibrations.

The twelve-week course was completed by twenty-one participants in the treatment group and fifteen participants in the placebo group and both groups were not significantly different at baseline in age or motor symptom severity. The authors found that 40Hz vibration administered by the special chair (which literally induces uniform vibration throughout the body, as distinct from WBV platforms that induce vibration only via the feet) was found to significantly reduce overall motor symptoms of Parkinson's disease (Mosabbir et al., 2020). The authors suggested that the most viable mechanism for the efficacy of low frequency vibration is that it may act to disrupt the pathological oscillatory activity within basal ganglia-thalamocortical circuits. The vibration transmitted throughout the entire body may act to perturb abnormally synchronised oscillations, in addition to eliciting supplementary releases of endogenous dopamine (Mosabbir et al.).

Neurogenic pain

Pain can also be experienced that is not a consequence of nociception, categorized as 'neurogenic' pain, stemming from neural circuit dysrhythmias or disconnections (Bartel & Mosabbir, 2021). However, neurogenic pain has been found to be mediated by vibratory analgesia as a result of cortical dynamics (Hollins et al., 2014). For example, in a study with fibromyalgia patients, positive effects were obtained due to oscillatory coherence, with 40Hz vibro-tactile stimulation of the body (Naghdi et al., 2015).

Another vibro-tactile study mentioned that patients with chronic pain syndromes, such as fibromyalgia, complain of widespread pain and tenderness, as well as non-refreshing sleep, cognitive dysfunction, and negative mood. Several lines of evidence implicate abnormalities of central pain processing as contributors for chronic pain, including dysfunctional descending pain inhibition (Staudl et al., 2011). Their study enrolled 28 normal pain-free controls, 29 fibromyalgia patients, and 19 subjects with neck or back pain. Pain was administered to all subjects via sensitivity-adjusted heat stimuli to each forearm and pain intensity was noted using a mechanical visual analogue scale (VAS). Subsequently, the 100Hz vibrating probe, with a round footplate (diameter 4cm) was placed either on the same forearm next to the heat stimulus or on the forearm opposite to the heat stimuli. The authors concluded that vibro-tactile stimulation effectively recruited analgesic mechanisms not only in normal pain free

https://doi.org/10.52202/078697-0007

controls, but also in patients with chronic musculoskeletal pain, including fibromyalgia (Staudl et al.).

The descending inhibition of pain by music and white noise

A second mechanism of pain mediation, sometimes referred to as the "top down" modulation of pain (Urien & Wang, 2019) but more accurately described as the "Descending Inhibitory system" (Pertovaara & Almeida, 2006) or "Descending Analgesia System" (Dobek et al., 2014) can be activated by music that creates a strong emotional response. Such music-invoked emotions can be described as "thrills" (Goldstein, 1980).

Music offers a wealth of benefits with no negative side effects and is, therefore, a favourable option for those who are looking for alternative pain management therapies (Pertovaara & Almeida).

The origin of this second pain mediation mechanism arose from an early study by Dr. Henry K. Beecher, titled "Pain in Men Wounded in Battle" in which he notes, "Three-quarters of badly wounded men, although they have received no morphine for a matter of hours have so little pain that they do not want pain relief medication... Strong emotion can block pain" (Beecher, 1946).

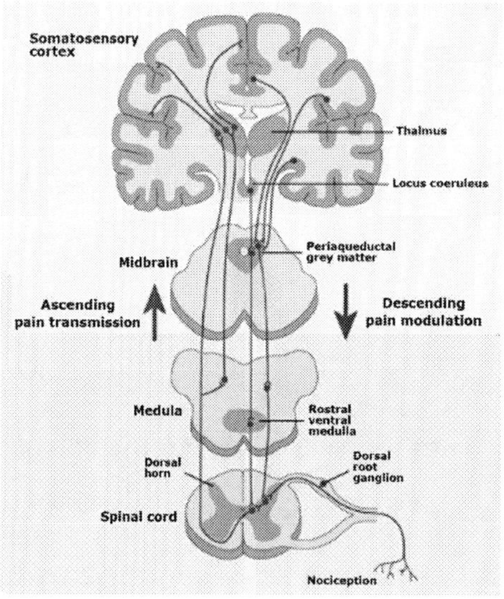

Ascending and Descending Pain Pathways
(Courtesy of walterskluwer.com)

Descending Inhibition concerns tracts arising from the brainstem that terminate on the spinal cord to suppress sensory transmission and consequently produce analgesia (Pertovaara & Almeida, 2006). Music induced analgesia is hypothesized to occur as a result of the release of opioids during music listening (Dobek, et al., 2014; Goldstein, 1980), thereby engaging the descending analgesia system that creates anti-nociceptive responses in the spinal cord. Descending inhibitory pathways use endogenous opioids, hydroxytryptamine (5-HT) and noradrenaline and their effects are mediated through supraspinal, midbrain-spinal and brainstem-spinal circuits (Suzuki et al., 2004). A large number of brainstem structures suppress pain through descending projections to the spinal dorsal horn, and in most cases their descending pain suppressive effect is relayed through the periaqueductal gray matter (PAG) and the rostral ventromedial medulla (RVM) (Pertovaara & Almeida, 2006). The RVM in the brainstem is a particularly important relay site for integrating descending influences to the spinal cord (Pertovaara & Almeida).

In an early (1960) seminal paper on this subject, titled "Suppression of Pain by Sound", by Gardner, Licklider and Weisz, the authors refer to "Audio Analgesia". They mention that in about 90% of 5,000 dental operations, including tooth extractions, sound stimulation was the only analgesic agent required (Gardner et al., 1960). Patients wore headphones and listened to relaxing music before the dental procedure. During the procedure the headphones were fed with white noise. The intensity of white noise used in these procedures was reported in another early paper (1962) "The Effects of White Sound and Music upon the Superficial Pain Threshold" by Robson and Davenport, to be as high as 120 decibels (Robson & Davenport, 1962). A foot switch was used by the dentist to lower the sound level to enable communication with the patient by an external microphone (Robson & Davenport). The reported 120 decibels was believed to "not damage hearing when used for relatively short periods," however, by today's safety standards this is a dangerous sound level and long or repeated exposure to sound at or above 85 dBA can cause hearing loss (A-weighted decibels)(Fink, 2017). However, the fact that so many dental operations were accomplished without the use of injected analgesics is impressive and suggests that trials of white noise at 85dBA should be conducted,

https://doi.org/10.52202/078697-0007

to assess if this safe level is efficacious as a descending analgesia substitute to injected analgesics, particularly for people who exhibit allergic reactions (Thyssen, 2008).

In the study: "Music Modulation of Pain Perception and Pain-Related Activity in the Brain, Brain Stem, and Spinal Cord: A Functional Magnetic Resonance Study," the authors concluded that when pain was administered [by a well-characterised thermal method] to female participants, who were asked to select their favourite music of any genre, perceived pain ratings were significantly reduced in activity in brain regions that process music and pain modulation (Dobek, et al., 2014). Brain stem and spinal cord structures showed similar activity changes concurrent with descending modulation, demonstrating activation of the PAG region with corresponding suppression within the dorsal horn region (Dobek, et al.). The authors also concluded that endogenous opioids may exert widespread analgesic effects by influencing several brain and brain stem regions within the descending analgesia pathway region (Dobek, et al.).

Breaking the "pain-spasm-pain" cycle in spinal injury, by sound

The first suggestion of a pain-spasm-pain cycle is generally credited to Janet Travell who wrote in 1942, "If muscle spasm causes pain, and pain reflexly produces muscle spasm, a self-perpetuating cycle might be established..." (Travell et al.,1942). Today, it is well known that spinal injuries typically create muscle spasm to "splint" the site of injury, providing protection while the healing process takes place.

In a round table discussion between four physicians, titled, "Diagnosis and Treatment of Low-Back Pain Because of Paraspinous Muscle Spasm: A Physician Roundtable," published in the journal *Pain Medicine*, Dr. McCarberg states:

> From an initial injury the patient develops pain. Motor neurons are activated as a reflex to splint that area causing muscle spasm. Muscle spasm clearly causes pain, but the exact cause of pain is poorly understood. Regardless, this pain will cause more muscle spasm...Hopefully, if this cycle is interrupted, a chronic problem will not occur. (McCarberg et al., 2011)

The trauma caused by a spinal injury, or other injury, causes pain, which leads to muscle tension. A cascade of effects then results in which the muscle tension decreases the blood circulation, which (hypothetically) causes hypoxia and further pain in the affected muscles. The

https://doi.org/10.52202/078697-0007

spasm then intensifies, which causes the hypoxia to intensify and the pain to intensify, therefore causing far more pain than the injury.

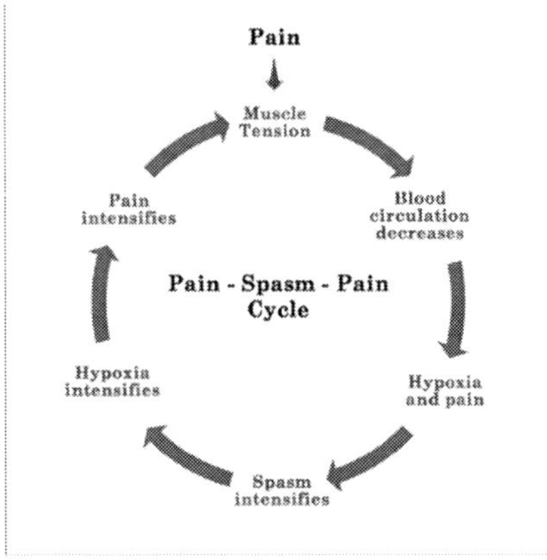

Pain > Muscle tension > Blood circulation decreases > Hypoxia and pain > Spasm intensifies > Hypoxia intensifies > Pain intensifies

The decrease in blood circulation is hypothesised to be a direct result of compression of intramuscular blood vessels, a concept that is supported by the fact that it is known that blood supply to a muscle is decreased during voluntary contraction and that pain following muscular exercise is very similar to pain induced by experimental reduction in the blood supply to a muscle (Roland, 1986). Further support for hypoxia in muscles comes from a study of intramuscular blood flow in the supraspinatus muscle of six healthy subjects. A muscle contraction pressure of 5.6 kPa (42mm Hg), which is 16% of maximal voluntary contraction, reduced local muscle blood flow significantly. The authors concluded that high intramuscular pressures found in supraspinatus muscle during exercise, impeded local muscle blood flow (Roland, 1986).

A potential solution to breaking the pain-spasm-pain cycle is by focally applied low frequency sound <50Hz. KKT International offer therapeutic support for spinal injuries by the application of very low frequency sound (primarily 16Hz), a non-invasive treatment to relieve pain by improving spinal alignment of the back, shoulders and pelvis, while enhancing the health of the spinal discs, ligaments and muscles (KKT International, 2023).

https://doi.org/10.52202/078697-0007

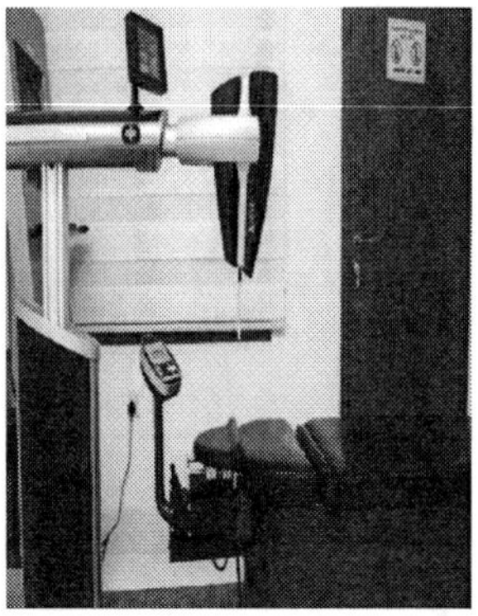

KKT Spine device

During the author's in vitro experiments with whole human blood, conducted in 2024, in collaboration with Professor Sungchul Ji of Rutgers University, significant increases in the level of pO_2 were measured with an Oxford Instruments "Oxylite" system when the blood was immersed in low frequency sound for 20-minutes. This research will be the subject of a forthcoming paper, however, the likely mechanism that underpins this phenomenon is the mechanical pressure generated by low frequency sound, similar to the low frequency pressure pulses from heart beats in the circulatory system, thus increasing oxygen-binding to haemoglobin molecules in red blood cells. Support for this potential method of increasing oxygen- binding, comes from the earlier mentioned study on Whole Body Vibration in which the oxygen saturation level was found to increase significantly between the first and second minute, following application of low frequency vibrations to supine patients (Boeselt et al., 2016).

Pain relief and anxiety relief by acupressure and sonopuncture

Acupressure is an alternative medicine methodology that originated in ancient China. Treatment effects are achieved by stimulating acupuncture points using acute pressure (Luo et al., 2013).

The World Health Authority, in their 1991 international acupuncture nomenclature report, lists 14 main meridians and 361 classical acupuncture points, in addition to 8 extra meridians and 48 extra points (World Health Organization, 1991). These same classical acupuncture points, which can be activated by acute local pressure, can also be activated by sound, since sound may be defined as: "Mechanical radiant energy that is transmitted by longitudinal pressure waves in a material..." (Merriam-Webster, n.d.). This is the basis of "sonopuncture," a therapeutic modality that is a species of acupressure.

A comprehensive review paper of fifteen acupressure studies concluded that acupressure is shown to reduce dysmenorrhea pain, labor pain, low back pain, chronic headache, and other traumatic pains. The clinical trials showed that acupressure can be efficiently conducted by health care professionals as an adjuvant therapy in general practice for pain relief (Chen & Wang, 2014). The authors also concluded that their systematic review paper begins to establish a credible evidence base for the use of acupressure in relieving pain and that an evidence-base of reliable and valid evaluation is crucial for clinicians. In terms of the implication for nursing education, practice, and research, the review provides important evidence that acupressure uses a noninvasive, timely, and effective way to support its effectiveness in relieving a variety of pains (Chen & Wang).

D. Carey, a licensed acupuncturist, developed a therapeutic method using tuning forks of specific frequency to activate acupuncture points while she was Clinical Dean at the Northwest Institute of Acupuncture and Oriental Medicine, in 1995. The intent was to seek a non-invasive therapy that could be taught to students and used in clinics with patient populations who were critically ill, including those suffering from HIV/AIDS, chronic pain, and trauma (Franklin, 2019). Today, this method of sonopuncture training is available in a certified course, providing an integrative medicine model that dovetails with many clinical specialities and can provide support for patients pursuing traditional Western medicine therapies (Franklin, 2019).

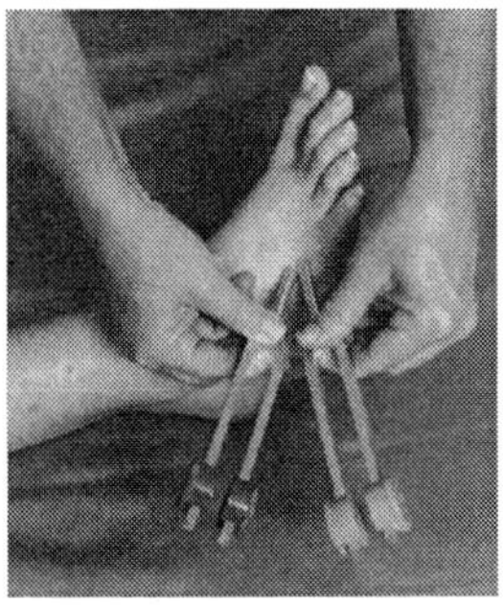

Sonopuncture, self-applied or practitioner-applied, by tuning forks, courtesy of
(Courtesy of Dr. E. Franklin)

Dr. E. Franklin, a colleague of D. Carey, who suffered from stress-related asthma, mild depression, and severe lower back pain, experienced immediate relief from one treatment with

https://doi.org/10.52202/078697-0007

tuning forks applied to acupuncture points. Her pain level was reduced from an 8/10 to a 3/10 and her depression lifted. Dr. Franklin submitted her dissertation in 2014, a multiple case study of self-care with seven hospital nurses, using tuning forks as an intervention to ameliorate symptoms of severe stress and compassion fatigue. At the outset of the study, six of the participants reported pain that they attributed to the stress in their lives. At the conclusion of the study, four reported reduced pain. All participants identified some aspects of working with the tuning forks that had archetypal significance and expressed interest in continuing to use the tuning forks. An important finding of the study was the recognition of the potential for tuning fork therapy in supporting patients suffering from pain and those in palliative care or hospice (Franklin, 2014).

Licensed acupuncturist, M.E. Wakefield, L.Ac., awarded "Educator of the Year" by the American Association of Oriental Medicine in 2005, is co-author of *Vibrational Acupuncture: Integrating Tuning Forks with Needles* (Wakefield & MichelAngelo 2020) with MichelAngelo, M.F.A., vibrational medicine advisor. Their book uniquely explores the synergy of tuning forks and acupuncture. For pain mediation, via sonopuncture, the authors recommend applying a tuning fork of 136.1Hz to specific acupuncture points. Although sonically activating acupressure points typically supports several interconnected bodily systems, the following examples focus mainly on pain mediation:

Lu-7 Lieque: "Broken Sequence" alleviates headaches, sore throat, migraines, toothache, pain in the wrist.

SI-3 Houxi, "Back Stream" alleviates neck pain, acute lumbar sprain, pain in the shoulder and elbow.

UB-62 Shenmai, "Extending Vessel" alleviates headache, backache, leg ache, insomnia.

TH-5 Waiguan, "Outer Pass" alleviates headache, facial pain, finger pain, hand tremors.

Bl-58 Feiyang "Taking Flight" mediates sciatic pain, alleviates headache, back pain.

Another important therapeutic use of tuning forks was discovered by E. D. McKusick, M.A., author of the book, *Tuning the Human Biofield* (McKusick, 2014). Energy-information is

https://doi.org/10.52202/078697-0007

constantly radiated from the body in the form of biofields, as mentioned in this article's introduction (Rubik, 2002). The biofields include bio-photonic energy, for example, modulated infrared electromagnetism that is a natural consequence of cellular metabolic processes, in addition to modulations in the electromagnetic fields emitted by the heart, brain and other organs. Quoting from the book's foreword by Dr. Karl H. Maret, who practices Complementary and Alternative Medicine,

> ...when a holographic sound field such as that produced by a tuning fork, which contains complex data structures of pure frequencies with changing phase relationships interacts with the biofield of a person, the cellular memories of various tissues can be reawakened, potentially leading to a healing response. Quantum physical field theory predicts the occurrence of a number of coherent dynamic phenomena in liquid water inside cells and tissues that may be stimulated by sound. This process affects the free electron clouds existing within these coherent water domains, [thus modifying] cellular processes through their interaction with the hydration shells surrounding cell membrane receptors (McKusik).

The biofield tuning method has been shown to consistently reduce anxiety, as well as relieve pain, and in a collaborative, grant funded, Institutional Review Board study, between the Biofield Tuning Institute and the Consciousness and Healing Initiative, fifteen participants with clinical anxiety received a one-hour session each week, for three weeks, followed by interviews. The study implemented the State-Trait Anxiety Inventory (STAI) to measure anxiety level, and participants were screened for clinically significant levels of anxiety using the Generalized Anxiety Disorder (GAD) system. All participants met clinical thresholds for generalized anxiety disorder. The results of the study were significant, with the mean level of STAI moving from 52 to 33. Most participants reported that although their anxiety did not completely go away, they were able to separate themselves from the anxiety and had developed a new relationship to it. Participants also found that they were more in touch with their anxiety triggers and, therefore, these triggers did not have as much of an impact or influence on their behaviour (Jain et al., 2023).

https://doi.org/10.52202/078697-0007

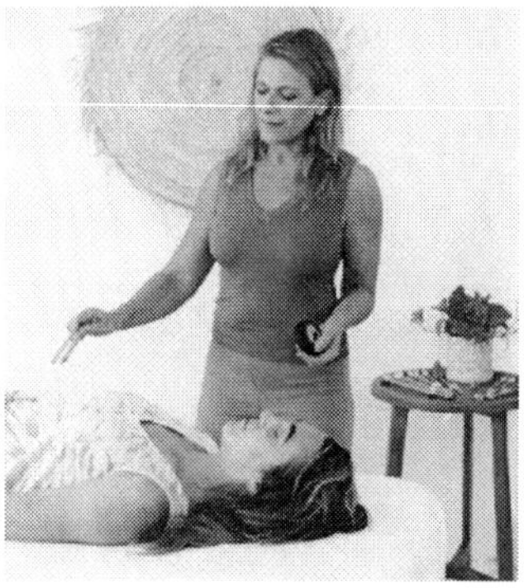

A biofield therapeutic tuning session in progress
(Courtesy of E. D. McKusick]

Pain relief by sonopuncture administered by electronic devices

Although sonopuncture is typically applied by tuning forks, devices that emit low frequency vibration can also achieve sonic activation of acupressure points (Novafon, 2024) in addition to devices that emit ultrasound (Progressive Acupuncture Wellness Center, n.d.). The acupressure points on the soles of the feet can also stimulate the meridian system by applying audible sound frequencies (Cyma Technologies, 2019). Dr. M. Cromwell developed a therapeutic device that uses a vibro-tactile transducer, emitting a range of audible sound frequencies into acoustic gel-filled pads on which the soles of the feet rest, thus stimulating the meridian system. Together with her assistant, Kate Holland, CCP, they conducted a six-week investigative pain study in 2016, with three individuals, a female of 30-years, a male of 38-years and a male of 68-years (International Sound Therapy Association, 2024).

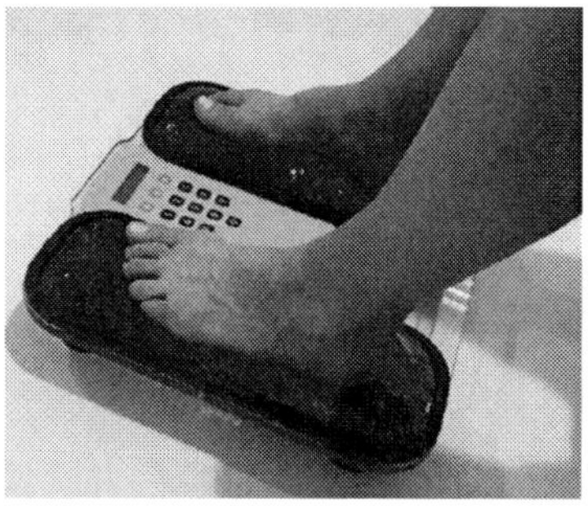

Sonopuncture applied via vibro-tactile transducer and acoustic gel-filled pads
(Courtesy of Dr. M. Cromwell)

https://doi.org/10.52202/078697-0007

The female participant entered the study with chronic headaches and sciatica and rated her neck pain at 6 on the VAS scale, which escalated to an 8 during headaches. She was unable to sleep due to her lower back pain and sciatica, which she remarked was constant and rated as a 6 on the VAS scale. She was taking over-the-counter medication for pain and insomnia. During the period of the study, the souls of her feet were sonically stimulated twice a week, for 30-minutes. During the third week she reported that her back pain had diminished to 3 on the VAS scale. By the end of the study, her ability to sleep had improved significantly and she was no longer taking over-the-counter pain medications. She also reported a reduction in the number and severity of headaches and rated the headache pain as 4 on the VAS scale (International Sound Therapy Association).

The 38-year old male entered the study with a rotator cuff injury of the right shoulder and was suffering chronic pain, registering 8 on the VAS pain scale, though not taking pain medication. His arm abduction was limited to 50 degrees. During the period of the study the souls of his feet were sonically stimulated once a week, for 30 minutes. During the sonic therapy he reported feeling warmth and movement of the muscles of his shoulders. At the end of the study, he reported minimal pain of 2 on the VAS scale and demonstrated that he had recovered a full range of motion of his right shoulder (International Sound Therapy Association).

The 68-year-old male military veteran entered the study with chronic neck pain, symptoms of PTSD and insomnia. He chose to participate in the study by utilizing a daily, at-home program, involving two thirty-minute daily sessions. He reported pain of 8 on the VAS scale and high levels of anxiety at varying times of the day or evening. He could only sleep two-three consecutive hours without waking and occasionally took over-the-counter sleep aids. After the first two weeks of daily sonic stimulation of the souls of his feet his sleep patterns improved and he decided to stop taking sleep medication. At the end of the six-week study he was able to sleep approximately 6 hours each night without interruption. At the end of the study the military veteran also reported a reduction of neck pain to 3 on the VAS scale and reported feeling less stress when going into public places (Cromwell & Holland, 2016).

https://doi.org/10.52202/078697-0007

In summary for this section on Acupressure/Sonopuncture, these studies show that there exists a significant potential for pain reduction, in addition to support for a range of other chronic conditions, including depression, PTSD, insomnia and others.

Oxygen-driven healing by sound

Limitations in the ability of the vasculature to deliver O_2-rich blood to tissue leads to, among other consequences, systemic hypoxia. Cells confronting hypoxic challenge either induce an adaptive response that includes increasing the rates of glycolysis and conserve energy, or undergo cell death (Chandan, 2009). Conversely, cells utilise O_2 in the aerobic metabolism of glucose to generate Adenosine Triphosphate (ATP) which fuels most active cellular processes, including tissue repair (Chandan).

Hypoxia sensing and response is activated upon exposure to a state of oxygenation that is lower than the pO_2 to which the cells or tissue is adjusted under normal conditions. This response cascade is centrally important in coping with the challenge of O_2 deficiency. Hypoxia is a hallmark of all ischemic diseases (Chandan). As mentioned earlier, in the section, 'Breaking the pain-spasm-pain cycle in spinal injury, by sound', pO_2 was shown to significantly increase during experiments with in vitro blood immersed in a low frequency sound environment, thus increasing oxygen binding to haemoglobin molecules in red blood cells. Support for this sonic method of increasing oxygen in blood, comes from the earlier mentioned study on Whole Body Vibration in which the oxygen saturation level was found to increase significantly between the first and second minute, following application of low frequency vibrations to supine patients (Boeselt, et al., 2016). Also, in the author's forthcoming paper on this subject, the optimal range of sonic frequencies, found during in vitro experiments to power the increase in pO_2 in blood, was 20Hz to 120Hz, which is similar to sounds created by human hearts (Reznekov, 1985) when auscultated by electronic stethoscope. This suggests that immersing patients in low frequency sounds is mimicking the heart's low frequency pressure pulses, thus supporting their pO_2 level.

Musical stimulation of the immune system (via headphones or full body immersion)

Illness in any form can cause emotional distress and emotions can play a significant role in a patient's recovery from illness or from an operating procedure. Stress and fear cause the release of cortisol [and adrenalin/ epinephrine] from the adrenal glands (Rosenberg, 2017) helping prepare the body for "fight or flight" by providing extra glucose, tapping into protein stores via gluconeogenesis in the liver (Thau et al., 2023).

However, cortisol also suppresses the immune system (Segerstrom & Miller, 2004) and other bodily systems considered by Nature to be "non-essential" in the short term, making the patient more vulnerable to contracting pathogens. While pharmaceutical sedatives are routinely prescribed to mediate a patient's stress and fear, music can produce a similar outcome without medication.

Music can evoke happy memories of times, places, or life events that can quickly transform a patient's mood into a sense of joy, in which state the brain and enteric nervous system in the digestive tract produces dopamine, which boosts the immune system (Salimpoor, et al., 2011; Papa, et al., 2017). In parallel with the increase in dopamine, a patient's favourite music causes a reduction in cortisol levels (Bartlett, et al., 1993). Joy also triggers the pituitary gland in the brain to release beta-endorphins into the bloodstream, which produce analgesia by binding to mu-opioid receptors that are present throughout peripheral nerves. Mu-opioid receptors have been identified in the central terminals of primary afferent neurons, peripheral sensory nerve fibres and dorsal root ganglia (Sprouse-Blum et al., 2010).

The pituitary gland also stores the neuropeptide, oxytocin, colloquially known as the "love hormone." Oxytocin is made in the hypothalmus and transported to large, dense-core vesicles of the posterior lobe of the pituitary gland (Bionity.com, n.d.) where it is released into the blood stream in response during sexual activity and orgasm in addition to childbirth. In a broader context there appears to be a general consensus among studies that music listening enhances oxytocin synthesis (Keeler et al., 2015) and postoperative patients listening to music through headphones demonstrated an increase in serum oxytocin and reported higher levels of relaxation, compared to a control group with no music (Nilsson, 2009). Oxytocin and its receptors appear to hold the leading position among the candidates for the substance of "happiness" (Magon & Kalra, 2011), and in a study focussed on autistic children, significantly lower levels of oxytocin was found in their blood plasma, suggesting a ray of hope in finding a role for oxytocin in treatment of autism (Magon & Kalra), that is, in both of these cases,

 https://doi.org/10.52202/078697-0007

(evoking happiness and supporting the treatment of autism) there is an obvious link in the form of music, whether applied via headphones or full body immersion.

A further important connection between music and the immune system was reported in a 2019 study by Augusta University, USA. The researchers found that when mice were subjected to low frequency sound vibrations, macrophages in their bloodstream proliferated significantly (Yu et al., 2019). This effect has not yet been demonstrated in humans; however, it seems likely that human blood will respond in a similar way to that of murine blood. The possible mechanism that powers the proliferation of macrophages in blood that is immersed in low frequency sound, is an increase in the pO_2 level, as discovered in the Whole Body Vibration study on supine humans, mentioned in the above section, 'Oxygen-driven healing by sound'. In that study the level of pO_2 was found to increase significantly between the first and second minute, following application of low frequency vibrations to supine patients (Boeselt, et al., 2016). It is important to mention that this aspect of the connection between music and the immune system would occur only during full body immersion, since the full circulatory system would require stimulation by low sonic frequencies.

Binaural Beats (via headphones) to create changes in brain state, with physiological benefits

Binaural beats were discovered accidentally in 1839 by the Prussian scientist, Heinrich Wilhelm Dove, during experiments with two tuning forks of dissimilar frequency. He has been referred to as "The Father of Meteorology" (Dove, 1879, p. 530) for his work in that field, however, as late as 1915 his discovery of binaural beats was considered a trivial special case of monaural beats (Oster, 1973). Monaural beats occur when two sounds of slightly different frequency sound simultaneously, resulting in a pulsating effect caused by the mixing of the two sounds, which are reinforced during moments when their phases align, and diminished when their phases oppose each other. But during headphone listening, when two slightly different frequencies are experienced, the composite difference frequency is known as a binaural beat and it provides a mechanism for stimulating the auditory system at very low frequencies, below the frequency range of hearing (Lane et al., 1998). Listening to binaural beats produces the illusion that the sounds are located somewhere within the head. The lower auditory centres of

https://doi.org/10.52202/078697-0007

the brain are in the medulla oblongata, and impulses from the right and the left ears first meet in the left or right superior olivary nucleus. These structures are part of the olive, an organ that in this view lies behind the brain stem. It is probable that binaural beats are detected here (Oster). The difference frequency between sounds presented to the left and right ears entrains the brain rhythms to that frequency.

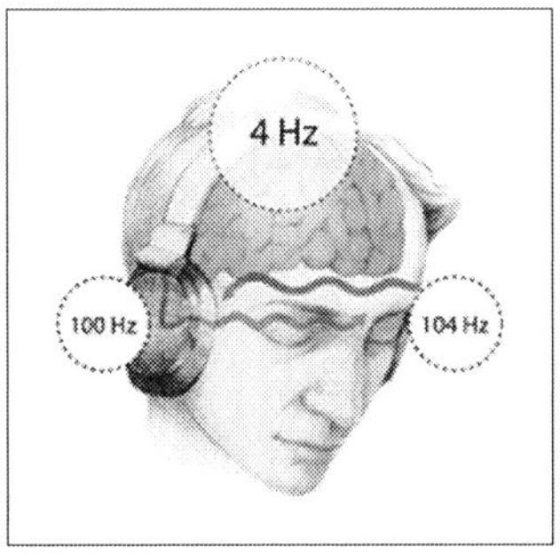

Depiction of binaural beats
(Courtesy of Hemi-Sync® hemi-sync.com)

In a carefully designed, double-blind, binaural beats crossover study, titled: "Binaural Auditory Beats Affect Vigilance Performance and Mood" in *Physiology and Behavior*, 29 volunteers were tested. The recordings used in the study contained a background sound of pink noise and a carrier tone, within which was embedded an entraining difference frequency between left and right channels. (The purpose of the pink noise was to mask the sound of the carrier tone.) The participants were kept blind to the true purpose of the study and were unaware of the presence of binaural beats in the headphones. The results of the study provided evidence that presentation of simple binaural auditory beat stimuli during a 30-minute vigilance task can affect both the task performance and changes in mood associated with the task. The effects on behaviour and mood were observed in the absence of participant expectations, and experimental control ruled out placebo effects. The authors concluded that simple binaural-beat auditory stimulation can influence psychomotor and affective processes, even when people are unaware that such signals are being presented, and that this technology may have applications for the control of attention and arousal and the enhancement of human performance (Lane et al., 1998).

https://doi.org/10.52202/078697-0007

In another double-blind crossover study, titled: "Reduced pain and analgesic use after acoustic binaural beats therapy in chronic pain - A double-blind randomized control cross-over trial," the authors concluded that theta rhythm binaural beats reduced pain intensity, stress and analgesic use, compared to sham stimulation, in chronic pain patients. A further conclusion was that the subsequent significant reduction in analgesic medication consumption in chronic pain patients' daily living could offer a valuable tool, augmenting the effect of existing pain therapies (Gkolias et al., 2020).

Robert Monroe of The Monroe Institute created a system of binaural beats in which individuals listen to a combination of audio binaural beats mixed with music, pink noise and/or the natural sound of the ocean waves that has been named the "Hemi-Sync" process. Studies with this system have shown improvements in sensory integration (Morris, 1990) relaxation, meditation, stress reduction, sleep, and pain management (Wilson, 1990) enriched learning environments and enhanced memory (Kennerly, 2013).

Sonic stimulation of the vagus nerve (via headphones) and by vocalisations

The vagus nerve system represents the main component of the parasympathetic nervous system, which oversees a vast array of crucial bodily functions, including control of mood, immune response, digestion and heart rate and carries an extensive range of signals from the digestive system and organs and vice versa (Breit et al., 2018). Upon exiting the jugular foramen, an auricular branch is given off, giving innervation to the auditory canal and external ear. This is the only branch of the vagus nerve given to the head. As the vagus nerve descends the neck via the medulla oblongata, branches leave to the pharynx and larynx before continuing into the thorax where it connects with the lungs, heart, and other major organs. The laryngeal and auricular connections are of special interest in the context of Sound Therapy and Music Medicine, discussed later in this section, following an overview of the vagus nerve system and methods of its therapeutic stimulation.

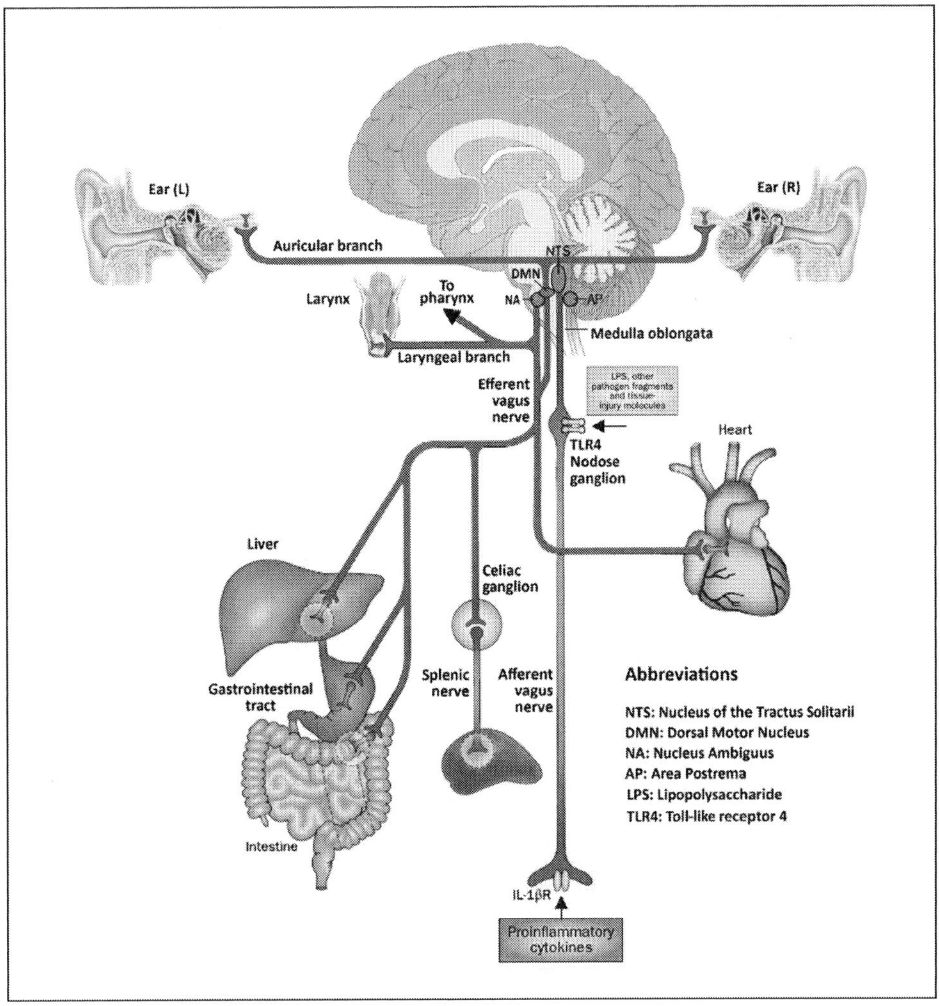

Vagus nerve branches and functional anatomy of the inflammatory reflex
(Adapted with permission, Pavlov and Tracey, 2012)

Schematic description:

Inflammatory mediators, such as cytokines, are released by activated macrophages and other immune cells upon immune challenge. These mediators are detected by sensory components of the afferent arm of the inflammatory reflex. Neural interconnections between the NTS, AP, DMN, NA and higher forebrain regions integrate afferent (red) and efferent (blue) vagus output, thus, regulating immune activation, supressing pro-inflammatory cytokines (Breit et al), and reducing inflammation. Vagus efferent output can be supported by auricular and laryngeal input.

https://doi.org/10.52202/078697-0007

The bi-directional communication between the brain and the gastrointestinal tract, sometimes called the "brain-gut axis," is a complex system that includes the vagus nerve system, and is becoming increasingly important as a therapeutic target for gastrointestinal and psychiatric disorders, such as inflammatory bowel disease, depression and posttraumatic stress disorder (Breit et al., 2018). The gut is an important control centre of the immune system, and the vagus nerves have immunomodulatory properties. As a result, this nerve system plays important roles in the relationship between the gut, the brain, and inflammation (Breit et al.).

There is a "hard-wired" connection between the nervous system and immune system as an anti-inflammatory mechanism. Counterregulatory mechanisms, such as immunologically competent cells and anti-inflammatory cytokines normally limit the acute inflammatory response and prevent the spread of inflammatory mediators into the bloodstream. The dorsal vagal complex responds to increased circulating amounts of tumour necrosis factor (TNF-α) by altering motor activity in the vagus nerves (Breit et al.), therefore, stimulation of the vagus nerves can help restore cytokine balance, leading to a reduction in chronic inflammation. The vagus nerve system is a major component of the neuroendocrine-immune axis which is involved in coordinated neural, behavioural, and endocrine responses that provide an important first-line innate defence against infection and inflammation and helps restore homeostasis in the body (Bonaz et al., 2016). Inflammatory diseases in which tumour necrosis factor (TNFα) is a key cytokine are good candidates for treatment targeting cholinergic anti-inflammatory pathway (CAP) (Bonaz et al.).

In essence, the inflammatory reflex is a physiological mechanism through which the vagus nerves regulate immune function and inhibit pro-inflammatory cytokine production (Pavlov & Tracey, 2012), thus, preventing excessive inflammation by alerting the brain to the presence of cytokines, which triggers the release of anti-inflammatory molecules that reduce the inflammation and maintain a healthy balance (Vagabond, 2024).

One of the most important potentials for vagal stimulation concerns its role in cancer prognosis. In a review paper titled, "The Role of the Vagus Nerve in Cancer Prognosis: A Systematic and Comprehensive Review," the authors highlight the fact that cancer remains the second leading cause of mortality worldwide, with prostate cancer being the most prevalent

https://doi.org/10.52202/078697-0007

cancer type in men and breast cancer in women. Cancer is a complex condition since it includes several hundreds of different types and because it involves and is affected by multiple body systems. Studies have shown that three basic biological factors contribute to the onset and progression of tumorigenesis: (1) oxidative stress, leading to DNA damage, (2) inflammation that contributes to escape from apoptosis, angiogenesis and metastasis, and (3) excessive sympathetic activity, which affects where cancer cells will metastasize. One factor common to these three factors, which inhibits all three and influences cancer prognosis, is vagal stimulation because it reduces oxidative stress, informs the brain about inflammation and profoundly inhibits inflammation, and inhibits sympathetic activity since it is a major branch of the parasympathetic nervous system (De Couck et al., 2018). An interesting and potentially crucial aspect of vagal activity concerns the link to Heart Rate Variability (HRV), the variability of interbeat cardiac intervals that is strongly correlated with vagal activity and cardiac autonomic regulation.

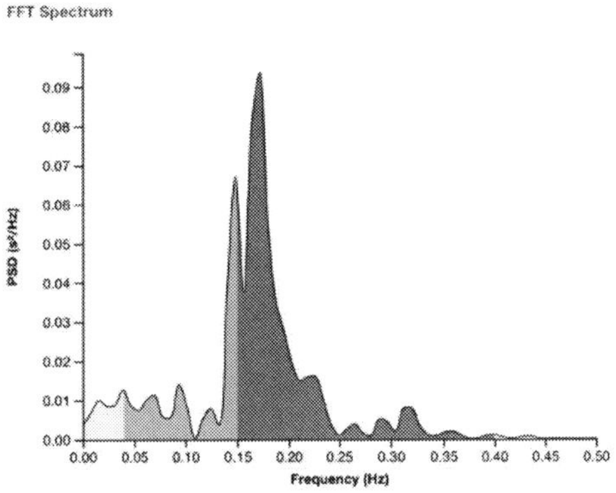

Power Spectral Density of HRV is typically expressed in milliseconds squared (ms^2), plotted against frequency
(Courtesy of EliteHRV.com)

The Power Spectral Density (PSD) of high frequency heart rate variability, (HF-HRV is cardiac frequency activity in the range, 0.15 to 0.40 Hz) is strongly associated with cardiovagal activity (Heathers, 2014). (By comparison, low frequency cardiac activity (LF) is in the range 0.04 to 0.15Hz). LF and HF frequency bands are widely used to quantify parasympathetic and sympathetic regulation (Heathers).

https://doi.org/10.52202/078697-0007

The vagus nerve system plays a major homeostatic role, indicated by people with high HRV who have shown improved recovery rates to physiological stress in cardiac, hormonal and immune systems, compared to those with lower HRV. In twelve studies that investigated the association between vagal tone activity and prediction of prognosis in cancer, which included 1822 patients, the emerging evidence was consistent in demonstrating a prognostic role of vagal activity and a significant correlation between survival time and high frequency heart rate variability. Using the vagal nerve index of HF-HRV, when data was analysed from a cohort of women with metastatic and recurrent breast cancer it was found that in a sample of 87 women, higher HF-HRV significantly predicted long term survival. It was also found that the predictive validity of HF-HRV improved when dividing it by the patients' heart rate, thus reflecting a more vagal/sympathetic ratio. The authors of the review study call for seriously considering adding HRV to the clinical estimation of prognosis in oncology (De Couck et al., 2018).

In the next section, methods of vagal stimulation are discussed including electrical, sonic, auricular and laryngeal, all of which improve vagal tone with many potential health benefits. The vagus nerve system can also be stimulated by acupuncture. Electrical stimulation of the vagus nerve system (VNS) was first studied in the 1930s and 1940s with animals, which laid the groundwork for studies in humans. Following successful clinical trials, the FDA approved the use of an implanted electrical vagus nerve stimulator for the treatment of certain types of epilepsy in 1997. The procedure involves implanting electrodes near the vagus nerve in the neck, along with a control device and battery implanted into the chest. The same mode of treatment was later also approved by the FDA for use in chronic, drug resistant depression (Vagabond, 2024).

Transcutaneous (through the skin) vagus nerve stimulation (tVNS) is currently emerging as an alternative and seeks to administer electrical stimulation to the vagus nerve system without the need for implant surgery, thus avoiding the associated risks. Stimulation is typically applied via the auricular branch of the vagus nerve via the tragus of the pinna. The European Union certified tVNS as an alternative treatment for epilepsy and pain in 2010 and 2012 respectively (Vagabond).

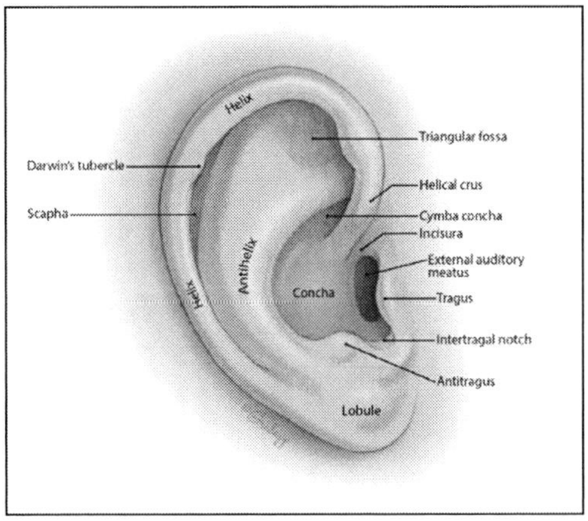

The pinna, showing the location of the tragus, where the vagus nerve terminates
(Courtesy of Tori Lewis Fibonacci Web Studio)

As early as 2001, researchers showed that electrical vagal stimulation via the tragus, using a form of electroacupuncture, reduced the dependence of patients with coronary arterial disease on vasodilator medication (Zamotrinsky et al., 2001). In their study, titled, "Vagal neurostimulation in patients with coronary artery disease," the authors stimulated the area of the ear near the auditory passage that contains endings of the auricularis nerve, by means of electrodes attached to short acupuncture needles, inserted to a depth of 0.1 to 0.3 mm. The authors concluded that electrical stimulation of the auricularis nerve results in tonic activation of the central vagus nerve structures and that an increase in vagal tone improves cardiac blood supply in patients with severe angina via dilation of spastic cardiac micro vessels (Zamotrinsky et al.). Referred pain to the ear from myocardial infarction has also been reported, due to the connectivity of ear and the heart, via the vagus nerve system (Amirhaeri & Spencer, 2010).

A carefully designed study concerning speech category learning in adults revealed that a simple, earbud-like device, developed at University of California San Francisco (UCSF), significantly improves the ability to learn the sounds of a new language. Mandarin Chinese is a particularly challenging language to learn due to its distinctive changes in pitch, which change the meaning of words. In their study, titled, "Non-invasive peripheral nerve stimulation selectively enhances speech category learning in adults" (Llanos et al., 2020), the authors moulded the shape of a participant's left ear using silicon putty. The resulting silicon

https://doi.org/10.52202/078697-0007

form was embedded with two-disc electrodes, 4mm diameter, contacting the cymba concha and cyma cavum of the outer ear, regions shown to be innervated by the auricular branch of the vagus nerve (Elliott et al., 2017). Electrical stimulation of the electrodes was provided by a commercial constant current device, deploying amplitudes below perceptual thresholds to allow participant blinding, resulting in stimulation intensities several milliamps lower than those used in previous non-invasive tVNS work (Llanos et al.).

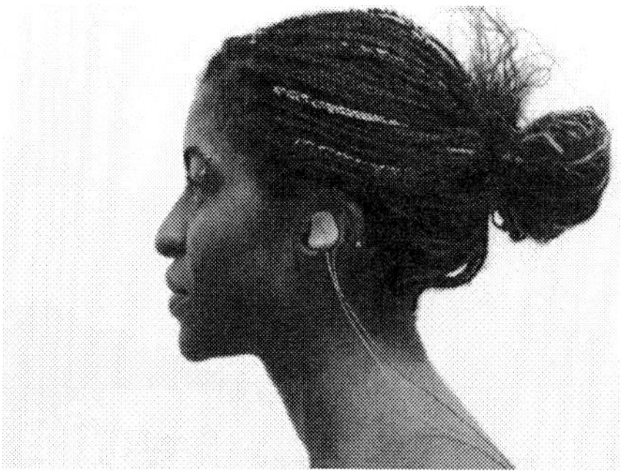

Experimental tVNS device, University of California San Francisco, Department of Neurological Surgery, Courtesy of Dr. M.K. Leonard

An earphone was inserted in the participant's right ear and fed with specific tones, to train the participant in tonal recognition and Mandarin speech training. The authors' primary focus was to investigate the extent to which pairing non-invasive, sub-perceptual threshold tVNS with behavioural training enhances the ability to categorize non-native speech categories in adults. Their results showed that when tVNS was paired with speech categories that were easier to learn, participants performed significantly better that those who did not receive tVNS stimulation (Llanos et al., 2020). Crucially, this group-specific learning improvement also generalised to new speech exemplars presented without accompanying stimulation and corrective feedback, thus demonstrating that tVNS can be used to accelerate speech perceptual learning in humans in a highly specific manner. The authors posit that sub-threshold tVNS engages the ascending brainstem network, and that some aspect of these neuromodulatory pathways cause a learning enhancement (Llanos et al.).

In the study titled "Anti-inflammatory properties of the vagus nerve: potential therapeutic implications of vagus nerve stimulation" (Bonaz et al., 2016), electrical tVNS frequencies used to activate vagal afferents to mediate depression and epilepsy are quoted as 20-30Hz and activation of the cholinergic anti-inflammatory pathway (CAP) as 1-10Hz. The authors mention anti-inflammatory properties of the vagus nerve system both through its afferent (activation of the HPA axis) and efferent (activation of the CAP) fibres and that it is a good therapeutic target in inflammatory conditions of the digestive tract, for example, irritable bowel syndrome, and rheumatoid arthritis.

Sonopuncture tVNS

Several commercial manufacturers are now producing devices that provide electrical transcutaneous vagus nerve stimulation (Electrocore, n.d.; Neuvanna, n.d.; Parasym, n.d.) and others that utilise infrasonic sound (Sensate, 2024). Returning to the subject of sonopuncture and to the research by the Niels Bohr Institute, discussed earlier in this article, it was shown that nerves conduct sound (soliton impulses), which in turn generate electrical pulses, due to the piezo- electric effect (Heimburg & Jackson, 2005). Therefore, although a corpus of research shows that the vagus nerve system can be electrically stimulated via the tragus and other acupressure points of the ears, it is clear that this can also be achieved sonically, and that such sonic stimulation will, automatically, lead to electrical stimulation of the auricular branch of the vagus nerve, due to the piezo-electric effect. In this scenario full ear headphones should be worn, enabling the full pinna of the ear to receive sound frequencies.

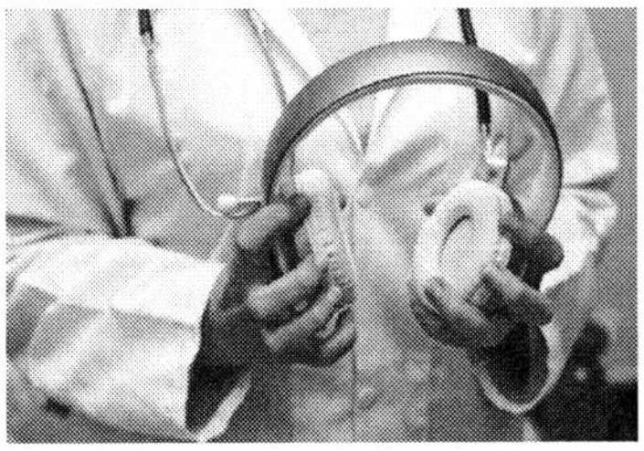

Very low frequencies for tVNS can be delivered sonically, via headphones

The very low frequencies commonly deployed in tVNS therapies can be created sonically via high specification headphones, and several manufacturers now produce

https://doi.org/10.52202/078697-0007

headphones that can deliver sounds as low as 5Hz (Beyer Dynamic, n.d.; Focal, n.d.; Senhheiser, n.d.). Although no studies of this type have yet been conducted, this form of sonopuncture may hold great therapeutic potential in support of a wide variety of illnesses, some of which have been mentioned in this section, including chronic inflammation. Sonic stimulation of the vagus nerve system could be achieved by sinusoidal tones, generated by an audio signal generator and fed to headphones via a suitable audio amplifier with the ability to handle very low frequencies. However, specially prepared music could also be deployed therapeutically, that is, music to which the very low frequencies identified in tVNS studies could be added to the music, either embedded in the recording or added separately to the amplifier input feed from an electronic signal generator. In such a scenario the patient would be able to enjoy the many health marker benefits of listening to music, mentioned earlier in this article, while the vagus nerve system would be vibrationally stimulated by sonopuncture frequencies below the range of hearing, adding further health benefits, for example, reducing chronic inflammation.

Before discussing vocal vagal stimulation, we make a special mention of the work of French otolaryngologist, Alfred A. Tomatis (b.1920, d. 2001). Dr. Tomatis received his Doctorate in Medicine from the Paris School of Medicine and formulated a theory that many vocal problems are actually hearing problems, based on the concept that *the voice cannot produce what the ear cannot not hear* [emphasis added], today referred to as "The Tomatis Effect."

Tomatis developed the "Electronic Ear," a device that utilizes bone conduction and sound filters to improve the tone of the muscles in the middle ear, to sensitize the listener to the missing frequencies, particularly in the high registers. The ear starts forming a few days after conception and is fully developed by the fourth month of pregnancy. Tomatis theorized that information coming from the fetal ear stimulates and guides the development of the brain. He believed that a number of auditory communication problems begin in pregnancy, with the fetus not properly responding to the voice of the mother. In children with ASD, he believed that his electronic ear device simulated the sound of the mother's voice as heard in the uterus, leading the child to gradually accept and respond to her real unfiltered voice. He reported that

https://doi.org/10.52202/078697-0007

this method often brought startling results, with children crying with joy as they recognized their mother's voice for the first time. He wrote:

> It is this [vagus] nerve that helps the singer to consciously rediscover the correct respiration rhythm as well as cardiac and visceral rhythms so that a synergy is created between this internal network and the larynx...It is equally important in mastering a fluid and correct verbal flow of speech...Without doubt singing is one of the best ways to free ourselves from the burden of parasympathetic or neurological imbalances" (Tomatis, 2005).

Vocal vagal stimulation

Last in this section, the laryngeal connection to the vagus nerves express and directly influence internal visceral states through the voice. In the article, "Stalking the calm buzz: how the polyvagal theory links stage presence, mammalian evolution, and the root of the vocal nerve" (Cazden, 2017) Joanna Cazden discusses Stephen W. Porges "polyvagal theory" which emphasises phonation, respiration and hearing. Porge's research proposes that the voice is strongly influenced by neuro-regulation that underlies our ability to communicate, and because the vagus nerve system mediates both our emotional state and our laryngeal muscle activity, our visceral states directly influence and are expressed through the voice.

Full appreciation of the autonomic vagus nerve, its influence on behaviour and its implications for vocal performance, requires a distinction between the neurophysiologic aspects of the autonomic system's two main sub-branches, the sympathetic and parasympathetic (Cazden). These two aspects of the autonomic nervous system can be thought of as a sympathetic accelerator and a parasympathetic break, providing bidirectional neural communication between our organs and brainstem (Siegel, 2012). Several nerve tracts in the brain can send sympathetic signals to stimulate a faster heart beat but only the vagus nerve sends a slowing signal, achieved during exhalation: the heart beats slightly faster as we inhale and slower as we exhale (Porges, 2017). This effect is termed, Respiratory Sinus Arrhythmia (RSA), which is a measure of vagal tone. The auditory nerve (CN VIII) that carries sound signals from the ears to the brain receives close crosstalk from the myelinated vagus nerve system. Porge mentions that the voice is a potent trigger of the physiological states of others and that emotional prosody is an audible sign of autonomic status, recognised in the brain of the listener. Because the laryngeal nerves branch directly from the

https://doi.org/10.52202/078697-0007

vagus, the voice transmits our inner resilience and expressive visceral state to others through sound (Cazden).

In the study, "Music Structure Determines Heart Rate Variability of Singers" (Vickhoff et al., 2013), it is suggested that singing can be viewed as initiating the work of a vagal pump: Singing produces slow, regular, and deep respiration which in turn triggers RSA, causing a pulsating vagal activity. In addition, as discussed in the section "Active and passive sonic stimulation of the nasal cavities and lungs," singing, chanting, and humming stimulates nitric oxide production in the nasal cavities and lungs, with many associated health benefits.

Playwright, John Guare, said, "the purpose of art is to exercise the muscles of the soul, so that when the challenges of life come, we are prepared." Porge's polyvagal theory suggests that these "muscles of the soul" may be found in the tiny area of the brainstem where a single myelinated pathway influences the remarkable vagus nerve system (Cazden, 2017).

These are just some of the many advances in medical science that hold the potential to support humankind in the quest to reverse disease, extend life and improve quality of life. The role of sound in medical modalities is growing each year for drug-free therapies and for diagnostic applications, and is finding welcome support among many physicians, and in hospitals worldwide. It seems increasingly likely that Sound Therapy and Music Medicine will have an important role in the future of medicine, one that deserves to be developed and nurtured.

https://doi.org/10.52202/078697-0007

References

American Music Therapy Association. (n.d.). Retrieved from https://www.musictherapy.org/about/

Amirhaeri, S., & Spencer, D. (2010). Myocardial infarction with unusual presentation of otalgia: a case report. *International Journal of Emergency Medicine*, *3*(4), 459–460. doi.org/10.1007/s12245-010-0222-8

Aristotle. (400 BCT-300 BCT/1995). *Poetics*. Loeb Classical Library. Cambridge: MA. p. 395. doi: 10.4159/DLCL.aristotle-poetics.1995

Ashutosh, K., Phadke, K., Jackson, J. F., & Steele, D. (2000). Use of nitric oxide inhalation in chronic obstructive pulmonary disease. *Thorax Journal, 55*(2). doi: 10.1136/thorax.55.2.109

Baldini, A., Von Korff, M., & Lin, E. H. (2012). A review of potential adverse effects of long-term opioid therapy: A practitioner's guide. *The Primary Care Companion for CNS Disorders*, *14*(3), PCC.11m01326. doi.org/10.4088/PCC.11m01326

Bartel, L. & Mosabbir, A. (2021). Possible mechanisms for the effects of sound vibration on human health. *Healthcare (Basil, Switzerland), 9*(5) 597. doi: 10.3390/healthcare9050597

Bartlett, D., Kaufman, D. & Smeltekop, R. (1993). The effects of music listening and perceived sensory experiences on the immune system measured by interleukin-1 and cortisol. *Journal of Music Therapy, 30*(4), 194-209. doi: 10.1093/jmt/30.4.194

Beecher, H. K. (1946). Pain in men wounded in battle. *Annals of Surgery,* 123(1): 96–105.

Beyer Dynamic (n.d.). model Amiron. 5Hz – 40,000Hz. Website. Retrieved from https://www.beyerdynamic

Bionity.com. (n.d.). Oxytocin. Retrieved from https://www.bionity.com/en/encyclopedia/Oxytocin.html

Boeselt, T., Nell, C., Kehr, K., Holland, A., Dresel, M., Greulich, T., Tackenberg, B., Kenn, K., Boeder, J., Klapdor, B., Kirschbaum, A., Vogelmeier, C., Alter, P., & Koczulla, A. R. (2016). Whole-body vibration therapy in intensive care patients: A feasibility and safety study. *Journal of Rehabilitation Medicine*, *48*(3), 316–321. doi: org/10.2340/16501977-2052

 https://doi.org/10.52202/078697-0007

Bogomolov, A. V., Dragan, S. P., & Erofeev, G. G. (2019). Mathematical model of sound absorption by lungs with acoustic stimulation of the respiratory system. *Dokl Biochem Biophys, 487,* 247–250. doi: org/10.1134/S160767291904001X

Bonaz, B., Sinniger, V., & Pellissier, S. (2016). Anti-inflammatory properties of the vagus nerve: Potential therapeutic implications of vagus nerve stimulation. The *Journal of Physiology, 594*(20), 5781-5790. doi: org/10.1113/JP271539

Bonica, J. J. (1953). *The management of pain*. Philadelphia: PA: Lea & Febiger.

Bradt, J., Dileo, C., & Potvin, N. (2013). Music for stress and anxiety reduction in coronary heart disease patients. *Cochrane Library, 12.* doi: org/10.1002/14651858.CD006577.pub3

Bradt, J., Potvin, N., Kesslick, A., Shim, M., Radl, D., Schriver, E., Gracely, E. J., & Komarnicky-Kocher, L. T. (2015). The impact of music therapy versus music medicine on psychological outcomes and pain in cancer patients: a mixed methods study. *Supportive Care in Cancer, 23*(5), 1261–1271. doi:10.1007/s00520-014-2478-7

Breit, S., Kupferberg, A., Rogler, G., & Hasler, G. (2018). Vagus nerve as modulator of the brain-gut axis in psychiatric and inflammatory disorders. *Frontiers in Psychiatry, 9,* 44. doi: org/10.3389/fpsyt.2018.00044

Britannica, Editors of Encyclopaedia. (2024). White noise. *Encyclopedia Britannica*. Retrieved from https://www.britannica.com/art/white-noise-music

Cantor, M. (1897). The life of Pythagoras. Open Court Magazine 6(493), 321-332.

Cazden, J. (2017). Stalking the calm buzz: How the polyvagal theory links stage presence, mammalian evolution, and the root of the vocal nerve. *Voice and Speech Review, 11*(2), 1-22. doi: 10.1080/23268263.2017.1390036

Chandan, K. S. (2009). Wound healing essentials: Let there be oxygen. *Wound Repair and Regeneration Journal, 17*(1), 1-18. doi: 10.1111/j.1524-475X.2008.00436.x

Chen, Y. W., & Wang, H. H. (2014). The effectiveness of acupressure on relieving pain: a systematic review. *Pain Management Nursing: Official Journal of the American Society of Pain Management Nurses, 15*(2), 539–550. doi: org/10.1016/j.pmn.2012.12.005

Chetan, H. [trans.] (n.d.). *Bhagavad Gita.* Self-published.

College Sidekick (2024). Retrieved from https://courses.lumenlearning.com/physics/chapter/17-3-sound-intensity-and-sound-level/

Cockroft, J. R. (2005). Exploring vascular benefits of endothelium-derived nitric oxide. *American Journal of Hypertension, 18*(S6), 177S-183S. doi:10.1016/J.AMJHYPER.2005.09.001

Corpus Hermeticum. (trans. 1999). Trans. Salaman C,. Oyen D., & Wharton, Richmond, UK: W. D. Gerald Duckworth & Co. Ltd, 1999.

Cromwell, M., & Holland, K. (2016). Effects of therapeutic sound on physical and emotional pain. *International Sound Therapy Association* (2024). Retrieved from: https://istasounds.org/research/effects-of-therapeutic-sound-on-physical-and-emotional-pain

Cyma Technologies. (2019). Retrieved from http://cymatechnologies.com

De Couck, M., Caers, R., Spiegel, D., & Gidron, Y. (2018). The role of the vagus nerve in cancer prognosis: A systematic and a comprehensive review. *Journal of Oncology, 2018,* 1236787. doi: org/10.1155/2018/1236787

Dennis, L., McNair, J. B., & Kauffman, L. H. (Eds.). (2013). *The Mereon matrix: unity, perspective and paradox*. Beijing, China: Elsevier Insights.

Diatchenko, L., Fillingim, R. B., Smith, S. B., & Maixner, W. (2013). The phenotypic and genetic signatures of common musculoskeletal pain conditions. *Nature Reviews Rheumatology, 9,* 340-350. doi: 10.1038/nrrheum.2013.43

Dobek, C. E., Beynon, M. E., Bosma, R. L., & Stroman, P. W. (2014). Music modulation of pain perception and pain-related activity in the brain, brain stem, and spinal cord: a functional magnetic resonance imaging study. *The Journal of Pain, 15*(10), 1057–1068. doi: org/10.1016/j.jpain.2014.07.006C

Dove (1879). Heinrich Wilhelm Dove. *Nature, April 10,* 529-530. Retrieved from https://www.nature.com/articles/019529d0.pdf

Dubinsky, T. J., Khokhlova, T. D., Khokhlova, V., & Schade, G. R. (2020). Histotripsy: The Next Generation of High-Intensity Focused Ultrasound for Focal Prostate Cancer Therapy. *Journal of Ultrasound in Medicine: Official Journal of the American Institute of Ultrasound in Medicine, 39*(6), 1057–1067. doi: org/10.1002/jum.15191

Electrocore (n.d.). Website. Retrieved from https://www.electrocore.com

https://doi.org/10.52202/078697-0007

Electrotherapy on the Web (2024). Retrieved from http://www.electrotherapy.org/modality/ultrasound-therapy?highlight=overview#Therapeutic%20Ultrasound%20Thermal%20and%20Non%20Thermal%20Effects%20Overview

Elliott, J. H., Synnot, A., Turner, T., Simmonds, M., Akl, E. A., McDonald, S., Salanti, G., Meerpohl, J., MacLehose, H., Hilton, J., Tovey, D., Shemilt, I., Thomas, J., & Living Systematic Review Network (2017). Living systematic review: 1. Introduction-the why, what, when, and how. *Journal of Clinical Epidemiology, 91*, 23–30. doi:org/10.1016/j.jclinepi.2017.08.010

Fitz-Gibbon, J. (2020, November 18) Doctor plays grand piano for 10-year-old cancer patient during surgery. *New York Post.* Retrieved from https://www.nypost.com/2020/11/18/doctor-plays-piano-for-ten-year-old-cancer-patient/

Fink, D. J. (2017). What is a safe noise level for the public. *American Journal of Public Health, 107*(1), 44–45. doi: 10.2105/AJPH.2016.303527

Focal (n.d.). model Utopia. 5Hz – 50,000Hz. Website. Retrieved from https://www.focal.com

Franklin, E. F. (2019). Practitioner self-care: A study in the applied use of sound-vibrations with Acutonics sound-based therapeutics. *Oriental Medicine Journal.* Retrieved from (PDF) Practioner Self-Care: A Study in the Applied Use of Sound Vibration with Acutonics®Sound-Based Therapeutics | Ellen Franklin - Academia.edu

Franklin, E. F. (2014). *Acutonics(RTM) self-care program and stress: Multiple case study exploration of an intervention to ameliorate symptoms of severe stress and compassion fatigue in nurses.* (Doctoral dissertation, Saybrook University).

Fyfe, M. C., & Bullock, M. (1985). Therapeutic ultrasound: Some historical background and development in knowledge of its effect on healing. *The Australian Journal Physiotherapy, 31(*6)220-224. doi:10.1016/S0004-9514(14)60635-8

Gardner, W. J., Licklider, J. C. R., & Weisz, A. Z. (1960). Suppression of pain by sound. *Science, 132*(3418), 32-33. doi: 10.1126/science.132.3418.32

Gkolias, V., Amaniti, A., Triantafyllou, A., Papakonstantinou, P., Kartsidis, P., Paraskevopoulos, E., Bamidis, P. D., Hadjileontiadis, L., & Kouvelas, D. (2020). Reduced pain and analgesic use after acoustic binaural beats therapy in chronic pain - A double-blind randomized control cross-over trial. *European Journal of Pain, 24*(9), 1716-1729. doi: 10.1002/ejp.1615

Goldman, J. & Goldman A. (2017). *The humming effect: Sound healing for health & happiness.* Rochester, VT: Healing Arts Press.

Goldstein, A. (1980). Thrills in response to music and other stimuli. *Physiological Psychology, 8*(1), 126-129. doi: 10.3758/BF03326460

Heathers, J. A. J. (2014). Everything hertz: Methodological issues in short-term frequency-domain HRV. *Frontiers in Physiology, 5.* doi:10.3389/fphys.2014.00177

Heimburg, T., & Jackson, A. D. (2005). On soliton propagation in biomembranes and nerves. *Biophysics and Computational Biology, 102*(28), 9790-9795. doi: 10.1073/pnas.0503823102

Henderson, T. A., & Morries, L. D. (2015). Near-infrared photonic energy penetration: Can phototherapy effectively reach the human brain? *Neuropsychiatr Dis Treat, 11,* 2191-2208. doi.org/10.2147/NDT.S78182

Henry, B. & Royston, J. T. (2017). A multiscale analytical model of bronchial airway acoustics. *The Journal of the Acoustical Society of America*, 142, 1774-1783. doi: 10.1121/1.5005497

Hollins, M., McDermott, K., & Harper, D. E. (2014). How does vibration reduce pain? *Perception, 43*(1), 70-84. doi: 10.1068/p7637

Holosonics.com (n.d.). Retrieved from https://www.holosonics.com/

Iamblichus (trans. 1818). *Life of Pythagoras*. Thomas, T., trans. from Greek. Rochester, VT: Inner Traditions. [Note: Iamblichus was a Syrian philosopher who lived circa 325 to circa 245 BC; therefore, a period of 170 years separates Pythagoras' death from Iamblichus' birth.

International Sound Therapy Association. (2024). Retrieved from: https://istasounds.org/research/effects-of-therapeutic-sound-on-physical-and-emotional-pain/

The Institute for Functional Medicine. (2024). Retrieved from https://www.ifm.org/functional-medicine/

Jain, S., McKusick, E., Ciccone, L., Sprengel, M., & Ritenbaugh, C. (2023). Sound healing reduces generalized anxiety during the pandemic: A feasibility study. *Complementary Therapies in Medicine, 74*. doi: org/10.1016/j.ctim.2023.102947

Järvholm, U., Styf, J., Suurkula, M., & Herberts, P. (1988). Intramuscular pressure and muscle blood flow in supraspinatus. *European Journal of Applied Physiology and Occupational Physiology, 58*(3), 219–224. doi: org/10.1007/BF00417252

 https://doi.org/10.52202/078697-0007

Jones, A. M., Vanhatalo, A., Seals, D. R., Rossman, M. J., Piknova, B., & Jonvik, K. L.(2021). Dietary Nitrate and Nitric Oxide Metabolism: Mouth, Circulation, Skeletal Muscle, and Exercise Performance. *Medicine and Science in Sports and Exercise.* doi: 10.1249/MSS.0000000000002470

John Hopkins Center for Music & Medicine. (2024). Retrieved from Music as Medicine | Johns Hopkins Center for Music and Medicine (hopkinsmedicine.org)

Joost-Gaugier, C. L. (2007). *Measuring heaven: Pythagoras and his influence on thought and art in antiquity and the Middle Ages.* Ithica, NY: Cornell University Press.

KKT International. (2023). Company website. Retrieved from http://www.kktspine.com

Kaufmann, K. (1989). Action Potentials and Electrochemical Coupling in the Macroscopic Chiral Phospholipid Membrane. Caruaru, Brazil. Unpublished.

Keeler, J. R., Roth, E. A., Neuser, B. L., Spitsbergen, J. M., Waters, D. J., & Vianney, J. M. (2015). The neurochemistry and social flow of singing: Bonding and oxytocin. *Frontiers in Human Neuroscience, 9,* 518. doi: org/10.3389/fnhum.2015.00518

Kennerly, R. C. (2013). Research paper. Retrieved from https://hemi-sync.com/research-papers/an-empirical-investigation-into-the-effect-of-beta-frequency-binaural-beat-audio-signals-on-four-measures-of-human-memory/

Lane, J. D., Kasian, S. J., Owens, J. E., & Marsh, G. R. (1998). Binaural auditory beats affect vigilance performance and mood. *Physiology and Behavior, 63*(2), 249-252. doi: 10.1016/S0031-9384(97)00436-8

Lipton, B. (2005) *The Biology of Belief.* Austin, TX: Elite Books.

Llanos, F., McHaney, J. R., Schuerman, W. L., Yi, H. G., Leonard, M. K., & Chandrasekaran, B. (2020). Non-invasive peripheral nerve stimulation selectively enhances speech category learning in adults. *NPJ Science of Learning, 5*(1), 12. doi: 10.1038/s41539-020-0070-0

Loscalzo J. (2001). Nitric Oxide Insufficiency, Platelet Activation, and Arterial Thrombosis. *Circulation Research. 88,* 756-762. doi: 10.1161/hh0801.089861

Lundberg, J.O., & Weitzberg, E. (1999). Nasal nitric oxide in man. *Thorax Journal. 54*(10). doi: org/10.1136/thx.54.10.947

Luo, D., Wang, X., & He, J. (2013). A comparison between acute pressure block of the sciatic nerve and acupressure; methodology, analgesia, and mechanisms involved. *Journal of Pain Research, 6,* 589–593. doi: 10.2147/JPR.S47693

Magon, N., & Kalra, S. (2011). The orgasmic history of oxytocin: Love, lust, and labor. *Indian Journal of Endocrinology and Metabolism,15 Suppl 3*(Suppl3), S156–S161. doi: org/10.4103/2230-8210.84851 (Suppl3): S156-S161

Maiorana, A., O'Driscoll, G., Taylor, R., & Green, D. (2003). Exercise and the nitric oxide vasodilator system. *Sports Medicine (Auckland, N.Z.), 33*(14), 1013–1035. doi: 10.2165/00007256-200333140-00001

Maniscalco, M., Weitzberg, E., Sundberg, J., Sofia, M., & Lundberg, J. O. (2003). Assessment of nasal and sinus nitric oxide output using single-breath humming exhalations. *The European Respiratory Journal, 22*(2), 323–329. doi: org/10.1183/09031936.03.00017903

Manniche, L. (1991). *Music and Musicians in Ancient Egypt*. London: British Museum Press.

Marberger, M., Stakl, W., Hruby, W., & Kroiss, A. (1985). Late sequelae of ultrasonic lithotripsy of renal calculi. *The Journal of Urology, 133*(2), 170-173. doi:org/10.1016/S0022-5347(17)48868-9

Mark, J. (2016). Imhotep. *World History Encyclopedia.* Retrieved from https://www.worldhistory.org/imhotep/

Mason, T. J. (2011). Therapeutic ultrasound, an overview. *Journal of Ultrasonics Sonichemistry. 18*(4), 847-852. doi: 10.1016/j.ultsonch.2011.01.004

Masuda, S. (1992). Role of the maxillary sinus as a resonant cavity. *Nippon Jibiinkoka Gakkai Kaiho, 95*(1), 71–80. doi: org/10.3950/jibiinkoka.95.71S

McCarberg, B. H., Ruoff, G. E., Tenzer-Iglesias, P., & Weil, A. J. (2011). Diagnosis and treatment of low-back pain because of paraspinous muscle spasm: A physician roundtable. *Pain Medicine, 12*(suppl. 4), S119-S127. doi: 10.1111/j.526-4637.2011.01253.x

McGill Newsroom Institutional Communications. (2013). Retrieved from https://www.mcgill.ca/newsroom/channels/news/major-health-benefits-music-uncovered-225589

McKusick, E. D. (2014). *Tuning the human biofield: Healing with vibrational sound therapy.* Rochester, VT: Healing Arts Press.

Medsonix. (2022). Retrieved from http://www.medsonix.com

https://doi.org/10.52202/078697-0007

Melzack, R., & Wall, P. (1965). Pain mechanisms: A new theory. *Science, 150*(3699). doi: 10.1126/science.150.3699.971

Merriam-Webster. (n.d.). Sound. In *Merriam-Webster.com. dictionary.* Retrieved from https://www.merriam-webster.com/dictionary/sound

Merskey, H., & Bogduk, N. (1994). *Classification of chronic pain*. 2nd ed. Seattle: WA. IASP Press, p. 1.

Mills, S. E. E., Nicolson, K. P., & Smith, B. H. (2019). Chronic pain: a review of its epidemiology and associated factors in population-based studies. *British Journal of Anaesthesia, 123*(2), e273–e283. doi:org/10.1016/j.bja.2019.03.023

Moilanen E., & Vapaatalo, H. (1995). Nitric oxide in inflammation and immune response. *Annals of Medicine*. *27*(3), 359-367. doi:10.3109/07853899509002589

Morris, S. E. (1990). Hemi-Sync and the facilitation of sensory integration. *Hemi-Sync Journal, 4,* 5-6.

Mosabbir, A., Almeida, Q. J., & Ahonen, H. (2020). The effects of long-term 40-Hz physioacoustic vibrations on motor impairments in Parkinson's disease: A double-blinded randomized control trial. *Healthcare (Basel, Switzerland), 8*(2), 113. doi: org/10.3390/healthcare8020113

Naghdi L, Ahonen H, Macario P, & Bartel L. (2015). The effect of low-frequency sound stimulation on patients with fibromyalgia: a clinical study. *Pain Research & Management. 20*(1). doi: 10.1155/2015/375174

Napoli, N. (2020). Music as medicine? 30 minutes a day shows benefits after heart attack. *American College of Cardiology.* Retrieved from https://www.acc.org/about-acc/press-releases/2020/03/18/09/02/music-as-medicine-30-minutes-a-day-shows-benefits-after-heart-attack

Neuvanna (n.d.). Website. Retrieved from https://neuvanalife.com/pages/using-xen

Nicodemus, N. E., Becerra, S., Kuhn, T. P., Packham, H. R., Duncan, J., Mahdavi, K., Iovine, J., Kesari, S., Pereles, S., Whitney, M., Mamoun, M., Franc, D., Bystritsky, A., & Jordan, S. (2019). Focused transcranial ultrasound for treatment of neurodegenerative dementia. *Alzheimer's & Dementia (New York, N. Y.), 5,* 374–381. doi: org/10.1016/j.trci.2019.06.007

Nilsson, U. (2009). Soothing music can increase oxytocin levels during bed rest after open heart surgery: A randomised control trial. *Journal of Clinical Nursing, 18*(15), 2153-2161. doi: 10.1111/j.1365-2702.2008.02718.x

Novafon. (2022). Retrieved from https://novafon.com/de/historie

Novafon. (2024). Retrieved from https://novafon.com/us/products

Nunn, J. F. (1997). *Ancient Egyptian medicine*. London: British Museum Press, 111 & 122.

Oster G. (1973). Auditory beats in the brain. *Scientific American, 229*(4), 94–102. doi: org/10.1038/scientificamerican1073-94G

Papa, I., Saliba, D., Ponzoni, M., Bustamante, S., Canete, P. F., Gonzalez-Figueroa, P., McNamara, H. A., Valvo, S., Grimbaldeston, M., Sweet, R. A., Vohra, H., Cockburn, I. A., Meyer-Hermann, M., Dustin, M. L., Doglioni, C., & Vinuesa, C. J. (2017). T_{FH}-derived dopamine accelerates productive synapses in germinal centres. *Nature* 547, 318–323. doi: org/10.1038/nature23013

Parametric Equaliser. (2024). Retrieved from https://www.presonus.com/learn/technical-articles/What-Is-a-Parametric-Eq

Parasym (n.d.). Website. Retrieved from https://www.parasym.co

Pavlov, V. A., & Tracey, K. J. (2012). The vagus nerve and the inflammatory reflex--linking immunity and metabolism. *Nature Reviews. Endocrinology*, *8*(12), 743–754. doi: org/10.1038/nrendo.2012.189

Pertovaara, A., & Almeida, A. (2006). Chapter 13. Descending inhibitory systems. *Handbook of Clinical Neurology, 81,* 179–192. doi: org/10.1016/S0072-9752(06)80017-5

Pietropaoli, A. P., Perillo, I. B., Torres, A., Perkins, P. T., Frazier, L. M., Utell, M. J., Frampton, M. W., & Hyde, R. W. (1999). Simultaneous measurement of nitric oxide production by conducting and alveolar airways of humans. *American Physiological Society, 87*(4), 1532-1542. doi: org/10.1152/jappl.1999.87.4.1532

Planck, M. (1944). *The nature of matter.* [Speech in Florance, Italy]. Retrieved from www. ivet. web. id Layanan Informasi 17 Jam.Das Wesen der Materie (from Archiv zur Geschichte der Max-Planck-Gesellschaft, Abt. Va, Rep. 11 Planck, Nr. 1797)

Porges, S. W. (2017). *The pocket guide to the polyvagal theory*. New York, NY: W. W. Norton & Co.

Prentice, W. E. (1998). *Therapeutic Modalities for Allied Health Professionals.* New York, NY: McGraw-Hill.

Progressive Acupuncture Wellness Center (n.d.). Sonopuncture. Retrieved from https://www.acupuncturechicagoillinois.com/sonopuncture.html

 https://doi.org/10.52202/078697-0007

Raja, S. N., Carr, D. B., Cohen, M., Finnerup, N. B., Flor, H., Gibson, S., Keefe, F. J., Mogil, J. S., Ringkamp, M., Sluka, K. A., Song, X. J., Stevens, B., Sullivan, M. D., Tutelman, P. R., Ushida, T., & Vader, K. (2020). The revised International Association for the Study of Pain definition of pain: concepts, challenges, and compromises. *Pain, 161*(9), 1976–1982. doi: org/10.1097/j.pain.0000000000001939

Reguera, G. (2011). When microbial conversations get physical. *Trends in Microbiology, 19*(3), 105-11. doi:10.1016/j.tim.2010.12.007

Reddi, D., Curran, N., & Stephens, R. (2013). An introduction to pain pathways and mechanisms. *British Journal of Hospital Medicine (London, England: 2005), 74 Suppl 12,* C188–C191. doi: org/10.12968/hmed.2013.74.sup12.c188

Reznekov, L. (1985). Understanding heart sounds and murmurs, with an introduction to lung sounds. *JAMA, 254*(1), 124–125. doi:10.1001/jama.1985.03360010134047

Rizk, M., Witte, M. B., & Barbul, A. (2004). Nitric oxide and wound healing. *World Journal of Surgery.* doi: 10.1007/s00268-003-7396-7

Robson, J. G., & Davenport, H. T. (1962). The effects of white sound and music upon the superficial pain threshold. *Canadian Anaesthetists' Society Journal, 9,* 105-108. doi: 10.1007/BF03021250

Rodriguez-Raecke, R., Niemeier, A., Ihle, K., Ruether, W., & May, A. (2009). Brain gray matter decrease in chronic pain is the consequence and not the cause of pain. *Journal of Neuroscience, 29*(44), 13746–13750. doi: 10.1523/JNEUROSCI.3687-09.2009

Roland, M. O. (1986). A critical review of the evidence for a pain-spasm-pain cycle in spinal disorders. *Clinical Biomechanics (Bristol, Avon), 1*(2),102-109. doi: 10.1016/0268-0033(86)90085-9

Rosenberg, J. (2017). The effects of chronic fear on a person's health. Conference, *Neuroscience Education Institute (NEI) Congress.* Retrieved from https://www.ajmc.com/view/the-effects-of-chronic-fear-on-a-persons-health

Rubik, B. (2002). The biofield hypothesis: Its biophysical basis and role in medicine. *The Journal of Alternative and Complementary Medicine, 8*(6). doi:10.1089/10755530260511711

Salimpoor, V. N., Benovoy, M., Larcher, K., Dagher, A., & Zatorre, R. J. (2011). Anatomically distinct dopamine release during anticipation and experience of peak emotion to music. *Nature Neuroscience, 14*(2), 257–262. doi: org/10.1038/nn.2726

Sawynok, J. (2016).Adenosine receptor targets for pain. *Neuroscience, 338,* 1-8. doi: 10.1016/j.neuroscience.2015.10.031

Schwaller de Lubicz, R. A. (1982) *Sacred Science*. Rochester, VT: Inner Traditions Bear & Co.

Schwiening, C. J. (2012). A brief historical perspective: Hodgkin and Huxley. *The Journal of Physiology, 590*(11), 2539-2809. doi: 10.1113/jphysiol.2012.230458

Segerstrom, S. C., & Miller, G. E. (2004). Psychological stress and the human immune system: A meta-analytic study of 30 years of inquiry. *Psychological Bulletin, 130*(4), 601-630. doi: 10.1037/0033-2909.130.4.601

Sennheiser (n.d.). model HD 820. 6Hz – 48,000Hz. Website. Retrieved from https://www.sennheiser.com/en-us

Sensate (2024). Website. Retrieved from https://www.getsensate.com/pages/how-it-works

Sethi, S., Yin, J., & Anderson, P. K., (2014). Lung flute improves symptoms and health status in COPD with chronic bronchitis: A 26 week randomised controlled trial. *Clinical and Translational Medicine.* doi: 10.1186/s40169-014-0029-y

Shrivastava, S., Kang, K. H., & Schneider, M. F. (2018). Collision and annihilation of nonlinear sound waves and action potentials in interfaces. *The Royal Society Publishing Interface, 15.* doi: 10.1098/rsif.2017.0803

Shrivastava, S., & Schneider, M. F. (2014). Evidence for two-dimensional solitary sound waves in a lipid controlled interface and its implications for biological signalling. *The Royal Society Publishing Interface, 11.* doi: 10.1098/rsif.2014.0098

Siegel, D. (2012). *Pocket guide to interpersonal neurobiology: An integrative handbook of the mind.* New York, NY: W. W. Norton & Co. New York.

Sprouse-Blum, A. S., Smith, G., Sugai, D., & Parsa, F. D. (2010). Understanding endorphins and their importance in pain management. *Hawaii Medical Journal, 69*(3), 70-71. PMC3104618

Stanley, T. (1687/2010). Pythagoras: His life and teaching, a compendium of classical sources. (J. Wasserman, Ed.) Lake Worth, FL: Ibis Press.

Staudl, R., Robinson, M. E., Goldman, C. T., & Price, D. D. (2011). Attenuation of experimental pain by vibro-tactile stimulation in patients with chronic local or widespread musculoskeletal pain. *European Journal of Pain, 15*(8), 836–842. doi: 10.1016/j.ejpain.2011.01.011.

Suzuki, R., Rygh, L. J., & Dickenson, A. H. (2004). Bad news from the brain: descending 5-HT pathways that control spinal pain processing. *Trends in Pharmacological Sciences, 25*(12), 613–617. doi: org/10.1016/j.tips.2004.10.002

https://doi.org/10.52202/078697-0007

Thau, L., Gandhi, J., & Sharma, S. (2023). Physiology, Cortisol. *StatPearls Publishing LLC*. Treasure Island: FL. Retrieved from https://www.ncbi.nlm.nih.gov/books/NBK538239/

Thaut, M. H. (2015). Music as therapy in early history. *Progress in Brain Research*, *217*, 143-158. doi: 10.1016/bs.pbr.2014.11.025

Thyssen, J. P., Menné, T., Elberling, J., Plaschke, P., & Johansen, j. D. (2008). Hypersensitivity to local anaesthetics – update and proposal of evaluation algorithm. *Contact Dermatitis, 59*(2), 69-78. doi: org/10.1111/j.1600-0536.2008.01366.x

Tian, Y., Liu, Z., Tan, H., Hou, J., Wen, X., Yang , F., & Cheng , W. (2020). New Aspects of Ultrasound-Mediated Targeted Delivery and Therapy for Cancer. *International Journal of Nanomedicine*, *15*, 401-418. doi: org/10.2147/IJN.S201208

Toda, N., Ayajiki, K., & Okamura, T. (2009). Cerebral blood flow regulation by nitric oxide: recent advances. *Pharmacological reviews, 61*(1), 62–97. doi: org/10.1124/pr.108.000547

Tomatis, A. (2005). *The ear and the voice*. Lanham, MD: Scarecrow Press.

Travell, J., Rinzler, S., & Herman, M. (1942). Pain and disability of the shoulder and arm: Treatment by intramuscular infiltration with procaine hydrochloride. *JAMA, 120*(6), 417–422. doi:10.1001/jama.1942.02830410005002

Urien, L., & Wang, J. (2019). Top-down cortical control of acute and chronic pain. *Psychosomatic Medicine, 81*(9), 851-858. doi: 10.1097/PSY.0000000000000744

Vatansever, F., & Hamblin, M. R. (2012). Far infrared radiation (FIR): Its biological effects and medical applications: Ferne infrarotstrahlung: Biologische effekte und medizinische anwendungen. *Photonics & lasers in medicine, 1*(4), 255-266. doi:10.1515/plm-2012-0034

Vickhoff, B., Malmgren, H., Astrom, R., Nyberg, G., Ekstrom, S-R., Engwall, M., Snygg, J., Nilsson, M., & Jornsten, R. (2013). Music structure determines heart rate variability of singers. *Frontiers in Psychology, 4.* doi: 10.3389/fpsyg.2013.00334

Wakefield, M. E., & MichelAngelo (2020). Vibrational *acupuncture: integrating tuning forks with needles,* (Ch. 2). UK: Singing Dragon Press.

Wang, W., Wang, S., Lin, W., Li, X., Andersen, L. L., & Wang, Y. (2020). Efficacy of whole body vibration therapy on pain and functional ability in people with non-specific low back pain: a systematic review. *BMC Complement Medicine and Therapies*, *20*(158). doi: org/10.1186/s12906-020-02948-x

Wang, X. M., Hamza, M., Wu, T. X., & Dionne, R. A. (2009). Upregulation of IL-6, IL-8 and CCL2 gene expression after acute inflammation: Correlation to clinical pain. *Pain*, *142*(3), 275–283. doi: org/10.1016/j.pain.2009.02.001

Wilson, E. S. (1990). Preliminary study of the Hemi-Sync sleep processor. *Colorado Association for Psychophysiologic Research.*

World Health Organization. (1991). *A proposed standard international acupuncture nomenclature*. Geneva: Switzerland.

 https://doi.org/10.52202/078697-0007

REPORT

Sound As Healing Light Waves

Iasos

Abstract

This article will address music and healing. Since focusing on creating music for over 50 years, I would like to share what I have learned about the mechanics of exactly how sound and music can induce healing. I will be discussing in detail the primary factors: 1. Physical Resonance 2. Emotional Resonance 3. Intent 4. Rhythmic Dance Music 5 Belief, and 6. Sound as a Channel for Your Consciousness

Keywords: Sound, Healing

https://doi.org/10.52202/078697-0008

Introduction

For any healing modality, healing is not creating something new but merely realigning the body to its own already-existing Divine blueprint. Healing is re-establishing a dynamic state of balance out of a temporary state of imbalance. There are innumerable ways to do this, and using sound is one potent way to do this.

Physical Resonance

Sound can heal primarily through physical resonance and emotional resonance. Resonance means: If there are two energy systems capable of vibrating to the same frequency, if one of them starts vibrating, and if it is sufficiently near the other one, it will tend to get the other one also vibrating to this frequency. For example, if you tune two strings on a guitar to the same note (same frequency), if you pluck one of these strings, the other string will also start vibrating.

Physical resonance results from the vibrating air "shaking up the molecules" and causing the energy system to synchronize with itself into "standing waves," which, out of random chaos produce order, symmetry, coherency, and stability. Hans Jenny with his study of Cymatics shows innumerable photographs of beautifully symmetric shapes, formed by randomly arranged powder on a surface which is caused to vibrate by sound from a speaker attached to that surface. Sound can produce order out of randomness, by causing an energy system to synchronize with itself, producing standing waves.

Standing waves are inherently self-reinforcing and accordingly tend to maintain their existence. Consequently, they are a fundamental characteristic of any self-organizing system, such as your body. And this increased coherency is the direction of increased health, longevity, and greater life force. This means that sound can cause the various parts of your body to synchronize with each other, such as your heartbeat synchronizing with your breathing and with your brain waves. Any system that self-synchronizes into standing waves causes these waves to reinforce each other (sustained good health), whereas in a vibrating system without standing waves, the waves tend to randomly cancel each other out (disease & death).

Shaking up the molecules for sound-healing through physical resonance can occur, for example, by toning with your voice or using gongs or using Tibetan bowls that are rubbed in a circular fashion. A more esoteric example is dolphins that have been known to use their sonar to

https://doi.org/10.52202/078697-0008

"tune up" the chakras (subtle energy centers) of humans swimming with them. This type of sound healing - physical resonance - usually does not involve a complex blend of many instruments, such as an orchestra, but usually only one potent sound by itself.

Emotional Resonance

Emotional resonance means that music influences your feelings, and your feelings directly affect your health. To properly understand this, one must understand how different dimensions can influence each other. Then we can apply this understanding to your physical body, your emotional body, and your mental body, each of which is vibrating in a different dimension, or a different "range of frequencies." Because of resonance, two systems tuned to the same frequency can have an energy-transfer between them. A simple example of resonance is radio: When your radio dial is tuned to the same carrier frequency that a particular radio station is transmitting on, this resonance allows your radio to pick up and "hear" the program from that radio station.

Now resonance also applies to the Law of Octaves. An octave higher means precisely twice the frequency and an octave lower means precisely half that frequency. For example, if "A" on a piano is 440 cycles per second, then 880 cycles per second is the "A" that is one octave higher and 220 cycles per second is the "A" that is one octave lower. Resonance can and does occur between octaves, simply because their frequencies are precise whole-number multiples of each other: If two strings on a guitar are tuned to 440 and 880 cycles per second, plucking either string will cause the other string to also vibrate, thanks to Resonance with the Law of Octaves.

Keeping this in mind, we can see how each of your bodies (physical, emotional, mental) can influence each other. Now to state the obvious:

1. Sound & music influence your emotions.

This is an energy transfer from the physical plane to the emotional plane. Furthermore,

2. Your emotions influence your physical body.

This is an energy transfer from the emotional plane back to the physical plane. Therefore:

3. Harmonious music/sound ==> harmonious feelings ==> good health.

Furthermore,

4. Your emotions influence your thinking.

https://doi.org/10.52202/078697-0008

This is an energy transfer from the emotional plane to the mental plane. And ...

5. Your thinking influences your physical body.

This is an energy transfer from the mental plane to the physical plane. Specifically, your cells respond to the mental pictures you hold in your mind of *how* your body is *supposed* to behave. Your beliefs about how your body is supposed to be will have long-lasting effects in how your body actually does behave.

Therefore:

6. Harmonious music/sound => harmonious feelings => optimistic thoughts => good health.

So to summarize, sound healing through emotional resonance triggers a chain reaction between dimensions that begin as a physical dimension sound and, through the Law of Octaves, passes through your emotional body and mental body, and then both your emotional and mental bodies influence the state of health of your physical body:

7. Harmonious music/sound ===> harmonious feelings ====> good health.

and

8. Harmonious feelings ===> harmonious feelings ===> optimistic thoughts ===> good health.

Please do not infer from this that you should repress or suppress inharmonious feelings. When you find yourself in such a state, the worst thing you can do is repress or suppress your feelings. The healthiest thing to do is to express them, release them, clear them and be done with them (study children - they're great at this!) ... rather than sustaining a state of inharmonious emotions. It is the long-term sustaining of inharmonious emotions that provides a supportive environment for poor health. The hypothetical ideal of course would be to create your reality in such a way that inharmonious emotions seldom get triggered in the first place.

Here's what happens when you repress/suppress your feelings: Your emotions are an egg-shaped energy field larger than your physical body and surrounding and interpenetrating your physical body. Each emotion is usually localized in one particular part of this emotional energy field. (For example, anger might be localized in your solar plexus.) Often, a person may decide that their emotion is bad, shameful, or too painful, and consequently decide to not let themselves

https://doi.org/10.52202/078697-0008

feel it. This causes the area where this emotion exists to freeze-up or lock-up, and basically get stuck and not flow. Since the flow of prana or life-force is extremely interactive with the flowing of emotional energy, if the emotional energy freezes up in one particular area, then the flow of life force also freezes up in that area. This means that not enough life force is flowing into that area now. So whatever physical organs happen to occupy the same space where the emotional energy got stuck are now no longer receiving sufficient life force! If this condition persists, then predictably those physical organs will begin having serious health problems!

Intent

Let yourself feel the feelings! Get the emotions flowing! (emotion = energy in motion.) All diseases are psychosomatic (caused by the mind and feelings). The physical body is the realm of "effect." The realm of "cause" is your thoughts and feelings. You can think of your physical body as the accumulated repository or record of all your thoughts and feelings throughout your life. The Angels of Healing make an important distinction between a temporary healing and a permanent healing: For them, a temporary healing is when only the symptoms (effect) are removed, not the cause. Permanent healing is when the emotional and mental attitudes creating the disease (cause) are changed or corrected. You can temporarily remove the symptoms, but for a permanent cure, you must remove the cause, and the cause is always in the feelings and thinking - your emotional and mental attitudes.

Furthermore, new incoming thoughts must pass through your intellectual censor, where you decide to accept or reject these ideas. However, music is sneaky: Sound and music bypass any intellectual censor and totally saturate their influence on your emotions & feelings. This highlights the importance of taking dominion over your own world and always monitoring what music you are allowing into your emotional world.

And for those of you that actively create and use sound in your work, remember intent can modulate (ride on) a sound wave just as one water wave can ride on top of another larger wave. For example, the same vowel sound of "ah" can make a plant shrivel up or thrive, depending on the intent of the singer in that moment. Maximize your intention when using sound. Clearly and consciously intend the effects of the sounds you are creating or using and watch its impact blossom right before your eyes! And intention creates an effect - not only when you are creating sounds yourself, but even when you are re-playing sound recordings created by others!

Rhythmic Dance Music

One area of sound healing that is seldom regarded as such is rhythmic dance music! How so? Rhythmic dance music can perform 3 powerful healing functions:

1. It can charge the bottom 3 chakras or energy systems in the body - the base of the spine, the sex center, and the solar plexus center.
2. It can help ground the physical body.
3. It can vitalize the physical body by pumping energy into it.

Rhythm is actually an energy-pump through time. And if it is an effective rhythm (you know ... the kind where you have to move), then it is pumping a lot of energy! Effective rhythms tend to induce a resonance between the pulsing sound waves and the pulsing body waves. When the resonance is great, they have a hi-tech name for this synchronized pulsing together of the sound waves and your body waves - dancing! Since resonance allows for energy-transfer, in such a synchronized state, the sound of the music you are dancing to can pump energy and vitality into your body! ("I could have danced all night...") Hence, rhythm can provide nourishment to your body.

The trick, however, is to only choose dance music that is emotionally positive - otherwise you can be polluting your emotional body while vitalizing your physical body. Be discerning, only choose dance music that you sense to be emotionally positive and uplifting. Trust your intuitive guidance on this.

Another area of sound healing is sacred music. Here is how this works: Whatever you place your attention on, your attention is like an electronic beam that instantly connects you to the target of your attention, like dialing a particular phone number. Think of this beam of your attention as a *wire* on the mental plane, connecting two points. While this electronic beam of your attention is connecting you to your attention-target, three things are occurring:

1. Your energy flows out to the object of your attention.
2. The energy from the object of your attention is flowing back to you.

https://doi.org/10.52202/078697-0008

3. By being electronically linked up, you and this object of your attention begin to synchronize, and entrain, and resonate together as one system.

Therefore, whether your attention is focused on television reporting some horrible disaster or focused on a beautiful flower, you and the object of your attention immediately begin resonating together as one synchronized system, and you immediately begin absorbing the qualities of that reality!

Belief

Now when your attention is on the One Supreme Source of all Life everywhere - that which makes all of us alive - then you begin resonating with this Universal Energy. This means you begin synchronizing with the entire universe! And since *balance* is one of the prime qualities of the Universal Source of All Life, this means you are getting more in balance with the entire universe! And balance = healing! Attuning to the Universal Source, by any means, is the most healing thing you can do for yourself!

However, most humans have such a short attention span, even if their attention is on the one Universal Source, it cannot solidly stay focused there more than a few seconds. Here is where music can help: Music can perform the function of assisting you in attuning to the Universal First Cause and in keeping your attention focused there for an extended time period. Sacred music performs the function of facilitating your attunement to the Universal Source and sustaining your focus on the Universal Source, thereby causing you to become more healed by establishing a sustained alignment with the Infinite Eternal Source of all Life. Just think - for each second that your attention is focused on the Universal Source of all Life, there is a tremendous stream of healing balancing energy flowing into your body and energy fields!

For those readers that professionally use sound as a healing tool, keep in mind that there are endless, innumerable, countless ways that sound can be used for healing, and no one knows them all! Know that your mind is an automatic magnet for ideas in any field of reality that your mind specializes in. If your mind specializes in sound-healing, know that your mind is that mental magnet for picking up ideas for unique new ways sound can be used for healing. So follow your hunches! Manifest those ideas - however "far out" they may seem! Dare to implement those flashes! Enrich this planet with the sound-healing techniques and ideas that you are receiving.

https://doi.org/10.52202/078697-0008

Having discussed sound as a healing modality, let me discuss some other uses of sound and music that are dear to my heart. (I apologize in advance if this seems a bit too metaphysical for some of our readers.)

Sound As a Channel for Your Consciousness

By "tuning" the antenna of your consciousness, you can more readily dial-in and access specifically desired light realms. When sound helps you do this, the sound is functioning as a vibrational gateway to higher planes. Rather than the "radio" dial of your consciousness being stuck most of the time on only one station (this physical dimension), you can exercise your ability to access different realms (practice moving that dial). The music then functions as training wheels, helping you dial in different light realms. Once you familiarize yourself with these realms, you can then readily access them without the music. Furthermore, you can even use sound as a "wave guide" to harmonically align with the higher octaves of your own being - your Higher Self. Through attention, you can dial in and travel through the various higher octaves or "floors" of your own God/Goddess Self.

There are vast legions of beings, both in the angelic kingdom and in the elemental nature kingdom, whose specialty is The Musical Outpouring of Divinity. In the angelic kingdom there are fleets of music angels - ready and eager to instantly come where they are invoked and invited to grace and join in with any ceremony, adding their uniquely exquisite energies to those of the humans. Very often humans do not hear the angels and do not even sense them, but they are nevertheless emotionally moved by them - to the very core of their souls.

As our planet continues its dimensional upshift, there will be more and more ceremonies that consciously include and combine the angelic kingdom and the elemental nature kingdom with the human kingdom. Music can be the unifying force that resonates these three kingdoms together into a unified whole - for worship, adoration, and alignment with the One Universal Source that makes all of us alive.

In the next few years, there will be appearing many musician-light-workers specifically focusing on creating music for all these purposes. Avail yourself of their sounds. Use sound to accelerate your evolution, rather than passively allowing sound to have a sluggish effect on your growth. Take charge of your Life - what music you hear. Look for music that is coherent,

https://doi.org/10.52202/078697-0008

emotionally uplifting, and centering. Music that feels like it is coming from a loving heart. Ask your own Heart: How is this music affecting me - on all levels? Trust what your Heart tells you, and act on it. Choose and intend to effortlessly ride these sound waves gracefully upward towards your own expansions - on waves of light.

https://doi.org/10.52202/078697-0008

Contributors

Dr. David Akombo is an accomplished scholar and researcher who has dedicated his career to uncovering the profound impact of music on human health and well-being. His journey began in Kenya, where he earned a Bachelor's degree in Music Education, sparking a lifelong curiosity about the intersection of music, culture, and healing. His further inquiry through research and education inspired him for further studies American institutions including Point Loma Nazarene University in San Diego, where he obtained a Master of Education degree, and at Bowling Green State University in Ohio, where he earned a Master of Music degree in Ethnomusicology. Dr. Akombo went ton to complete his academic journey with a Doctor of Philosophy degree in Music Education from the University of Florida in Gainesville, Florida, a premier American institution. His rich academic experiences have shaped his research, which spans Africa, Indonesia, Japan, and the United States, culminating in groundbreaking contributions to the fields of music and medicine. Among his most notable achievements is the founding of the *Interdisciplinary Society for Quantitative Research in Music and Medicine* (ISQRMM), an organization dedicated to fostering collaboration and advancing the understanding of music's therapeutic power. Dr. Akombo currently serves as Distinguished Professor and Dean of the Faculty of Culture, Creative, and Performing Arts at the University of the West Indies in Barbados. He can be reached at dakombo@isqrmm.com

Antonios Chrysovergis (Tony Crisos) is a musician, writer, philosopher, and lecturer. He studied Music Performance with a special focus on Jazz at Berklee College of Music (BA) and received his Masters in Music Education from Boston University. He has studied ancient Greek Philosophy, as well as Western Esoteric traditions, and performs on both the guitar and the ancient Greek Lyre. Antonios, as a writer and philosopher, lectures on the Orphic tradition throughout the country and has published several articles on music and philosophy.
He can be contacted at tcrisos@me.com.

Adina Dabija, L.Ac., MSTOM, Dipl. O.M. is a writer, philosopher and western esoteric arts practitioner. Adina currently lives in New York, where she practices Chinese medicine. She published several books of poetry and fiction. She is also a priestess of Dionysus in the ancient Orphic mystery tradition, an initiatic lineage of Hellenic priesthood passed down directly from teacher to student. Adina can be contacted at imilovan@unipu.hr.

Marianne Gubri, harpist, composer, therapeutic musician, researcher in musicology, and music therapy, speaker, writer, and lecturer combines an original and unique musical journey with a quest for personal development. She has published more than 10 record albums, a book on historical music therapy *Una melodia per l'aere luminoso* (OM edizioni) and several courses on harp teaching. Director of the harp school Arpeggi, and of the Bologna Harp Festival, she speaks at TEDx conferences, international courses, conservatories and universities. A graduate in the United States in therapeutic music with IHTP, in VAHT Vibroacoustic Harp Therapy, and a Music and Imagery student, she is the Director of International Harp Therapy Program for France and Italy and organises seminars and events on harmonisation, resonance, and well-being through sound. She graduated in the Celtic harp in Brittany in 1997 afterwards she graduated in ancient harps at the Conservatoire of Tours, and she received a Master's in Musicology at the University of Venice /Tours in 2002. She performs in numerous concerts and international tours in Italy, France, Portugal, Ireland, Belgium, Switzerland, Sweden, Holland, Spain, Mexico, Brazil, USA, Dubai, China, India, Egypt with her own compositions. She accompanies her concerts with readings, meditations, and artistic research that take the audience into a wonderful inner world, beyond the threshold of appearances. She combines her concerts with conferences, seminars, book presentations and sound healing experiences, sharing insights and techniques learned in her 20 years' experience at the International School of Self Awareness. She has produced numerous shows combining different languages such as literature, theatre, art, dance, video, and integrating research with an appreciation of female creativity and spirituality. The author can be contacted at info@mariannegubri.com

Iasos (pronounced ya' sos) is a music-creator, specializing in heavenly, celestial, inter-dimensional music. He is also one of the original founders of "New Age" music, releasing his first album in 1975. Since then, he has remained active in creating celestial music and healing sounds. He can be contacted at iasos@iasos.com

Dr. Brian Manternach, is an associate professor (clinical) in the Department of Theatre at the University of Utah and a research associate at the Utah Center for Vocology, where he is on the faculty of the Summer Vocology Institute. His voice students have been cast in film, TV, national and international Broadway tours, regional Equity theatres, and cruise lines. Honors include the Distinguished Teaching Award from the University of Utah, Teacher of the Year Award from the Cal-Western Region of the National Association of Teachers of Singing (NATS), Faculty Excellence in Research Award from the University of Utah College of Fine Arts, and the Voice Pedagogy Award from the NATS Foundation. He has given presentations for the Pan American Vocology Association, Voice Foundation, Voice and Speech Trainers Association, Fall Voice Conference, NATS National Conference, Voice Study Centre, and TEDxSaltLakeCity, among others. He is an associate editor of the *Journal of Singing* and a columnist for *Classical Singer* and has been a contributing author to four books. Additional research, essays, and reviews have been published in the *Journal of Voice*, *Voice and Speech Review*, *Musical Theatre Educators' Alliance Journal*, and *VOICEPrints*, among others. His book, *Voices of Influence: Exploring the Journey of a Teacher and Student in the Voice Studio* (Rowman & Littlefield) will be published in 2025. Born and raised in Iowa, Dr. Manternach earned a B.A. in music from Saint John's University/College of Saint Benedict in Minnesota, a Master of Music from the University of Wisconsin-Milwaukee, and a Doctor of Music degree from the Indiana University Jacobs School of Music. brianmanternach.com. Brian can be contacted at brian.manternach@utah.edu

Dr. Jeremy Manternach is Associate Professor and Program Coordinator of music education at the University of Iowa. He teaches undergraduate and graduate choral pedagogy, voice science, music education, and research courses and coordinates and conducts the UI Youth Chorale, a high school outreach choir. Dr. Manternach was a nominee for the UI College of Education Collegiate Teaching Award and Iowa Faculty Scholar Award and was the 2021 recipient of the Early-Career Outstanding Researcher Award. Dr. Manternach has conducted state, conference, region, and city honor choirs in Arizona, Colorado, Idaho, Iowa, and Kansas. As a high school choral director in Minnesota, he received a district-wide Teacher Outstanding Performance Award and directed the Andover HS concert choir at the ACDA-MN state convention. Dr. Manternach's research interests include choral and vocal acoustics, conducting gesture and singer efficiency, adolescent voice, and pre-service teacher voice use. He frequently presents refereed research presentations and workshops at music education, choral, and voice science conferences. Journal publications to date include articles in the *Journal of Research in Music Education*, *Bulletin of the Council for Research in Music Education*, *Psychology of Music*, the *International Journal of Research in Choral Singing*, the *International Journal of Music Education*, the *Journal of Voice*, *Voice and Speech Review*, the *Journal of Music Therapy,* and the *Journal of Music Teacher Education*. He also recently appeared on the "Music and Medicine on Mondays" video series and the "Everyday Musicality" podcast. He can be contacted at jeremy-manternach@uiowa.edu.

Ana Paul is a Romanian-American 19-year old musician currently based in Long Island, New York. During her formative years she attended the German International School of New York in Westchester, where she learned German fluently and received her DSD-II certification, making her trilingual. She attended Fiorello H. LaGuardia High School as a vocal major and released her first album, *Lady Moon*, in 2020 at age 16. Ever since then she has been releasing music on streaming platforms each year, composing all of her own songs and lyrics, recording and producing several of them by herself. She also studied hypnosis and received her certification in 2021. She is currently studying Psychology on a pre-med track at the University at Buffalo. She can be reached at anapaul3756@gmail.com

Dr. John Stuart Reid is an acoustics engineer, scientist and inventor with a fascination for sound and electronics that go back to early childhood. At age 15 he built a radio telescope with which he listened to the signals of the early satellites orbiting overhead, and noise from the stars. He studied pure electronics at Northumbria University and went on to work under Professor Joseph Joshua Weiss and Dr Alastair Johnson in the Department of Radiation Chemistry where he assisted with their Febetron beta-ray field-emission device for which he built a range of high energy pulse experiments.

He also worked with an Electron Spin Resonance machine. Reid left the University to begin his own sound engineering company that eventually developed into a full acoustics consultancy business and included operating high-powered argon-ion lasers for the entertainment industry. In 1997 he conducted a series of experiments in Egypt's Great Pyramid to study its remarkable acoustics. He devised an experimental setup that would permit the resonances of the granite sarcophagus to be rendered visible, using cymatics principles. During the experiments he experienced a profound healing of his lower back that sparked his interest in sound's ability to heal. The Great Pyramid research was a great success and led him to conclude that making sound visible was potentially a powerful new investigative tool for science. He published the Great Pyramid research in an article titled, Egyptian Sonics. In the years since the Great Pyramid experiments, he has continued to develop cymatics technology, resulting today in the CymaScope, a new type of scientific instrument. It is now possible to image any sound—within the instrument's bandwidth—with a high degree of reproducibility. His cymatics research is helping to elevate this important new field in the scientific arena, and he speaks at scientific conferences in Europe and the USA. In the medical field his study of the effects of music on human blood, using the CymaScope instrument to differentiate between the sounds emitted by cancer cells and healthy cells was published in the Water Journal, and has far reaching health implications. In 2021 he was invited to write a chapter on Sound Therapy and Music Medicine for a new medical text book, which was published in 2022, in English, Russian, and Slovak. A free copy is available at: https://www.acuclinic.eu/ecompendium/.

Also, in 2022 he hosted a highly successful course with The Shift Network, teaching Cymatics, Sonic Science, and Music Medicine and in late 2022 and early 2023 he hosted a successful advanced course with the Shift Network, based on Frequency Medicine and Cymatic Science. Dr. Reid's study on how dolphins see with sound and communicate sono-pictorially, published in the Journal of Marine Science, carries the potential for eventual interspecies communication, a future in which we will be able to hold a conversation with a dolphin. His two recent studies are focused on the effects of music on the longevity of red blood cells, with a preliminary report published by Experiment.com, and on differentiating between the sounds emitted by healthy cells and cancer cells, a scientific paper published in the *Water Journal*. He can be reached at john@sonic-age.com.